Everyday Mathematics®

Student Math Journal 1

The University of Chicago
School Mathematics Project

Columbus, OH • Chicago, IL • Redmond, WA

The McGraw·Hill Companies

UCSMP Elementary Materials Component
Max Bell, Director

Authors
Max Bell
Jean Bell
John Bretzlauf*
Amy Dillard*
Robert Hartfield
Andy Isaacs*
James McBride, Director
Kathleen Pitvorec*
Peter Saecker

Technical Art
Diana Barrie*

**Second Edition only*

Photo Credits
Phil Martin/Photography, Jack Demuth/Photography, Cover Credits: Bee/Stephen Dalton/Photo Researchers Inc., Photo Collage: Herman Adler Design Group

www.sra4kids.com

Send all inquiries to:
SRA/McGraw-Hill
P.O. Box 812960
Chicago, IL 60681

Printed in the United States of America.

ISBN 0-07-584462-1

1 2 3 4 5 6 7 8 9 DBH 07 06 05 04 03 02

The McGraw·Hill Companies

Name Date Time

A Geometric Pattern

1. Color the pattern. Use 2 or more colors.

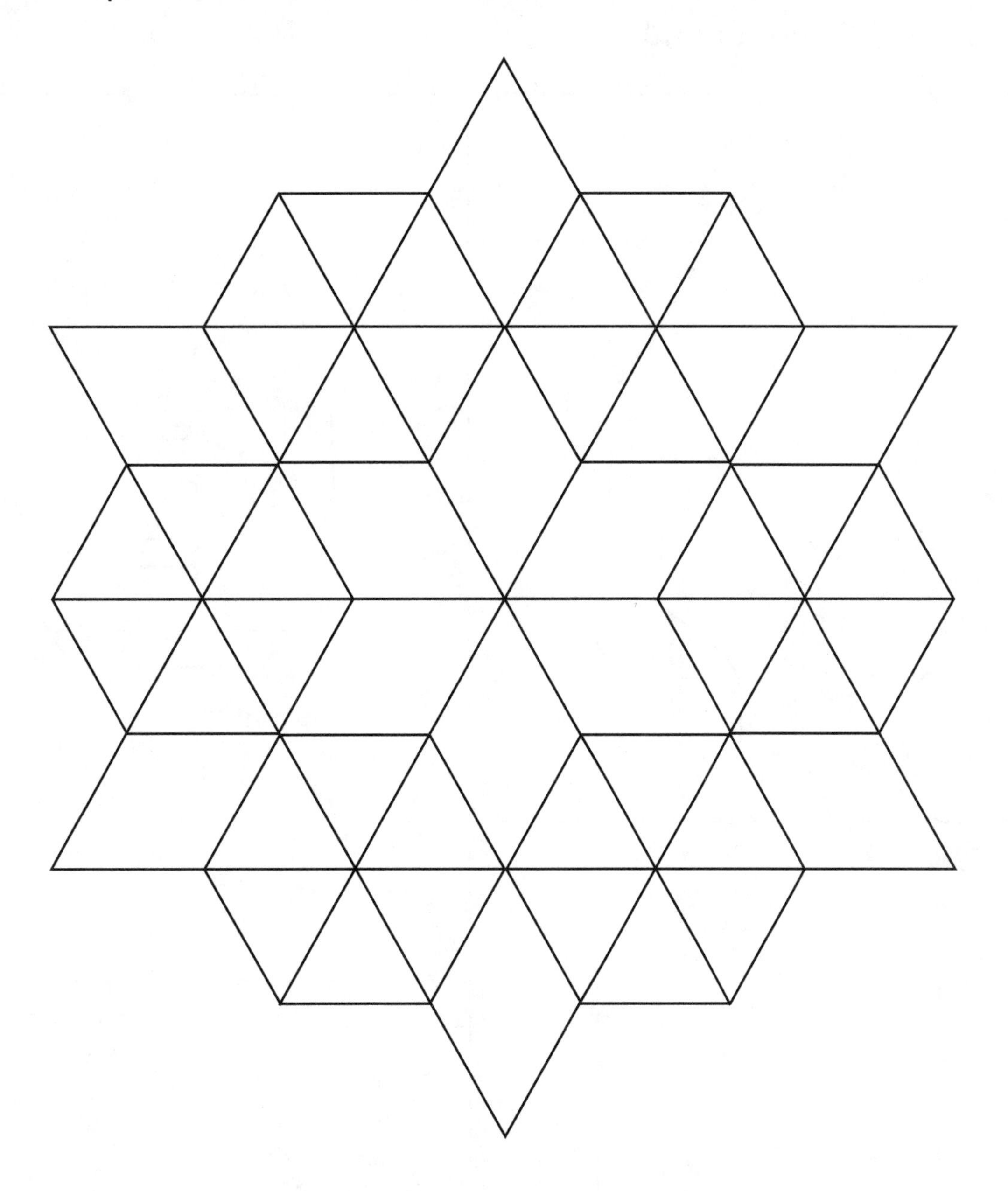

2. Trace your finger around triangles, rhombuses, trapezoids, or hexagons in your design.

Name Date Time

Lines of Symmetry

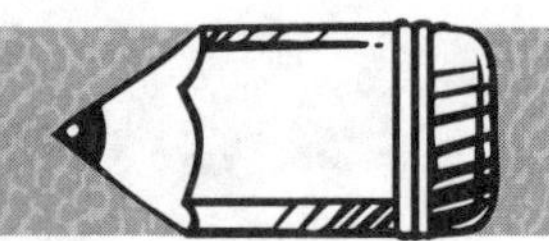

Cut out each shape. Find all the lines of symmetry for each shape by folding it in half.

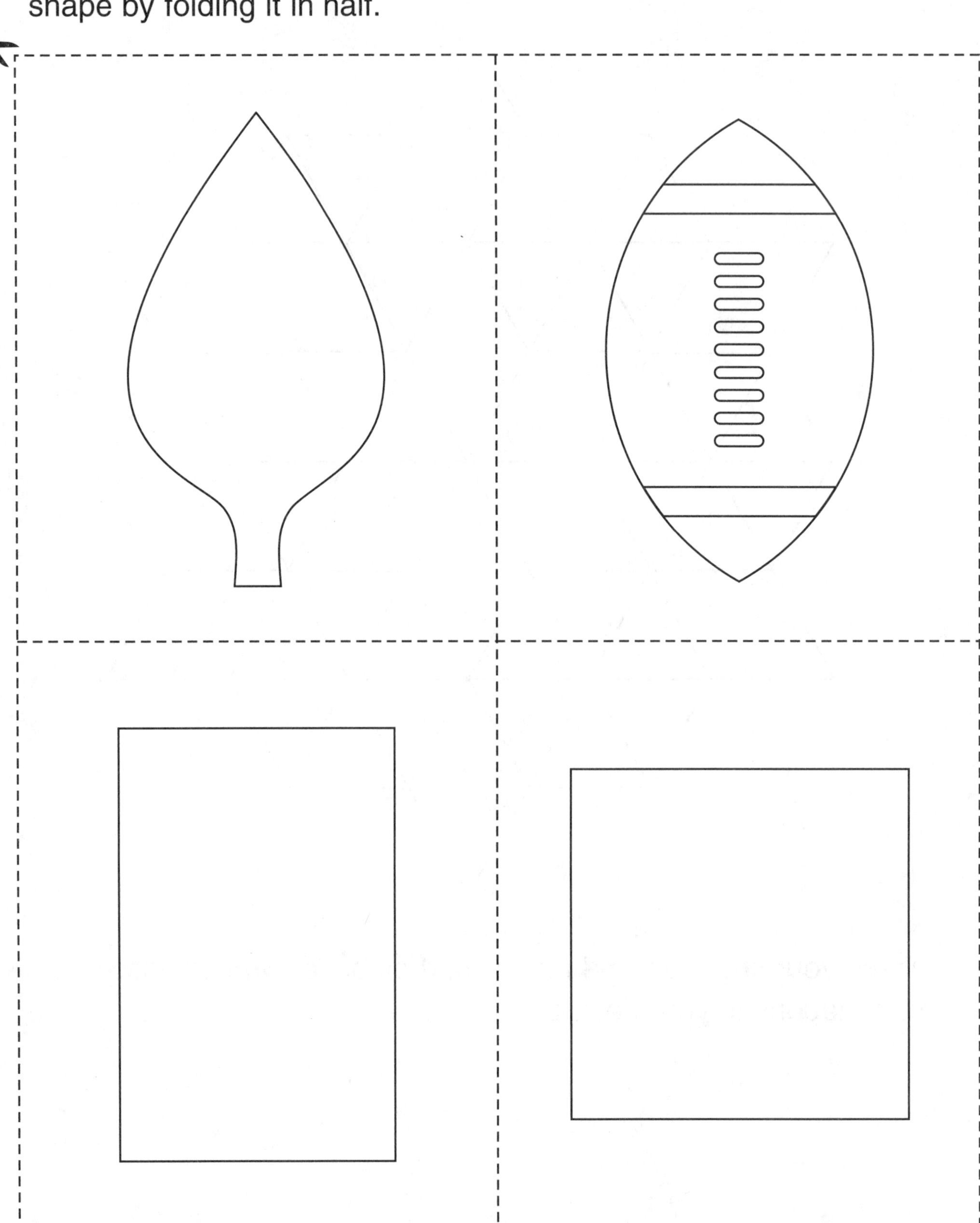

Contents

Unit 1: Numbers and Routines

A note at the bottom of each journal page indicates when that page is first used.
Some pages will be used again during the course of the year.

Unit 2: Addition and Subtraction Facts

Unit 3: Place Value, Money, and Time

Unit 4: Addition and Subtraction

Unit 5: 3-D and 2-D Shapes

Unit 6: Whole-Number Operations and Number Stories

Activity Sheets

Date Time

Number Sequences

Fill in the missing numbers.

1.

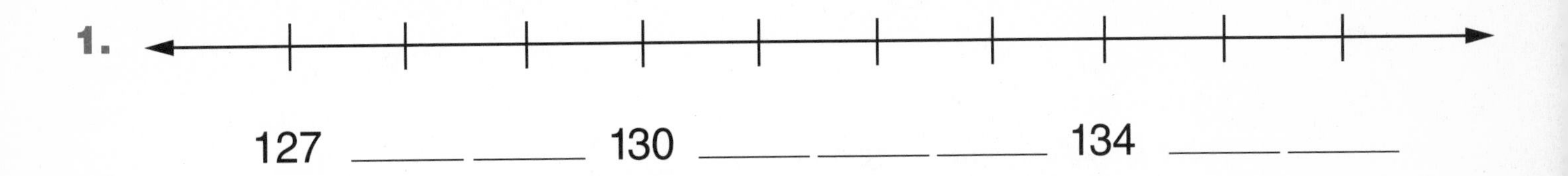

2.

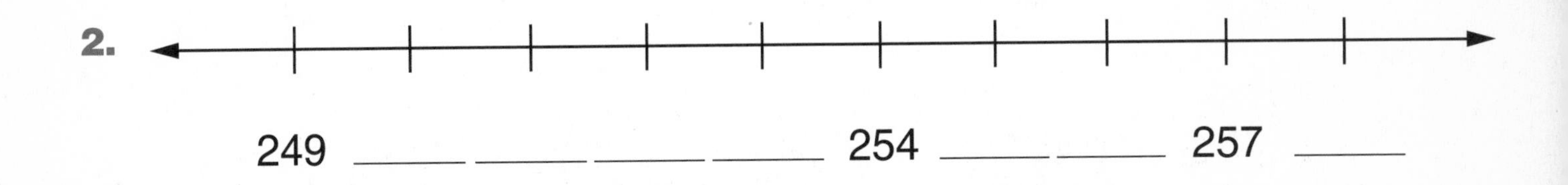

3.

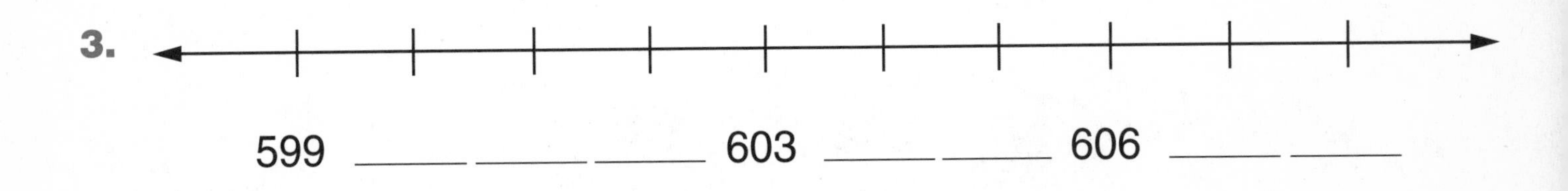

4.

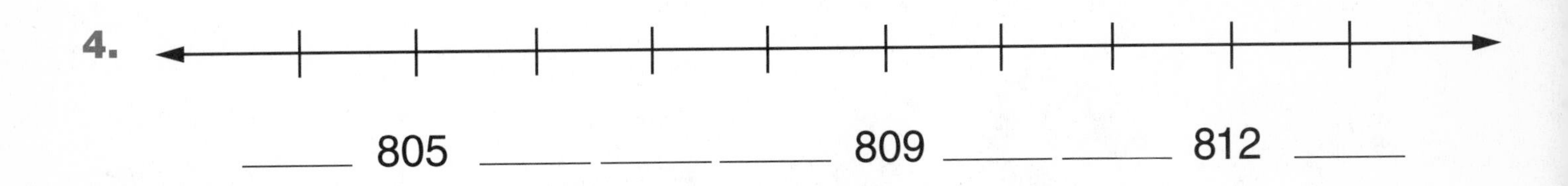

5.

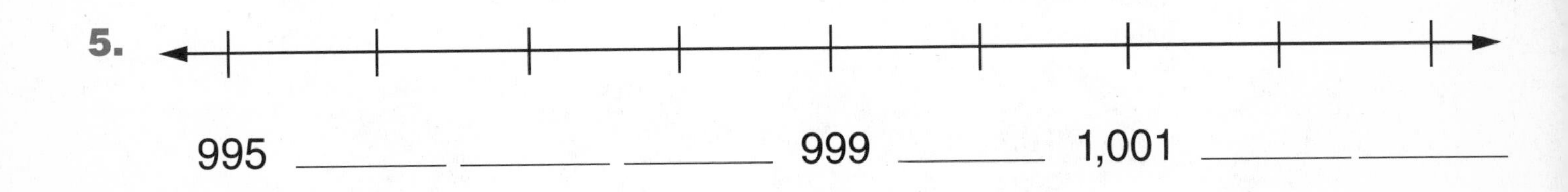

6.

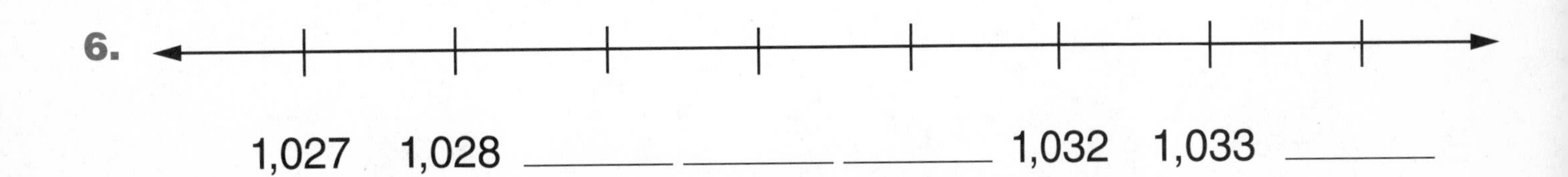

Date Time

Coins

1.      = ______¢

2.     = ______¢

3.      = ______¢

4.     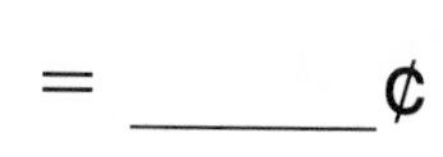 = ______¢

5.     = ______¢

6.      = ______¢

7.      = ______¢

Date Time

Calendar for the Month

Month ________

Sunday	Monday	Tuesday	Wednesday	Thursday	Friday	Saturday

Date Time

Time

Write the time.

1.

___ : ___

2.

___ : ___

3.

___ : ___

Draw the hands.

4.

5:00

5.

1:45

6.

9:30

Draw the hands and write the time.

7.

___ : ___

8.

___ : ___

9.

___ : ___

Use with Lesson 1.3.

Date Time

Addition Facts

1. 

$$\begin{array}{r} 2 \\ +\ 4 \\ \hline \square \end{array}$$

2. 

$$\begin{array}{r} 5 \\ +\ \square \\ \hline 9 \end{array}$$

3. 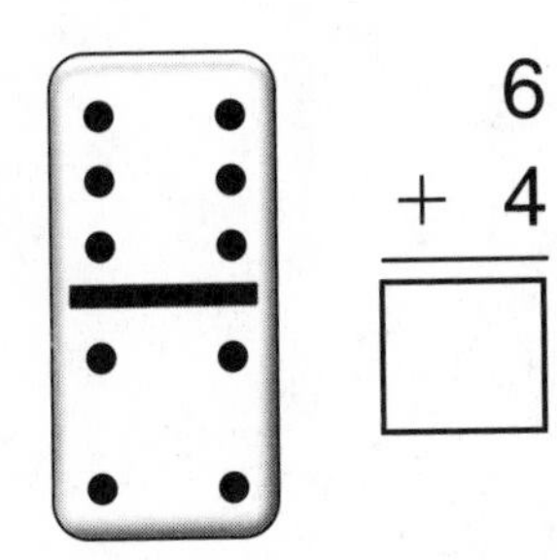

$$\begin{array}{r} 6 \\ +\ 4 \\ \hline \square \end{array}$$

4. 

$$\begin{array}{r} \square \\ +\ 9 \\ \hline 16 \end{array}$$

5. $6 + 8 = \square$

6. $8 + \square = 11$

7. $\square + 4 = 10$

Number Grid

8. Fill in the number grid.

				185					

Tallies

9. Write tally marks for 18. ______________

10. Write the number. 𝍸 𝍸 𝍸 𝍸 𝍸 𝍸 || ______

Date Time

Counting Bills

Write the amount.

1.

= $______

2.

= $______

3.

= $______

4.

= $______

Challenge

5.

= $______

Date Time

Math Boxes 1.7

1. Fill in the missing numbers.

15, 20, ______, ______,

______, 40

2. Write the time.

_____ : _____

3. Write the amount.

$______

4. Count by 2s.

18, ______, 22, ______,

______, ______

5. Write a 3-digit number.
Read it to yourself.

6. Write these numbers in order.
Start with the smallest number.

40 23 81

______ ______ ______

Date Time

Math Boxes 1.8

1. Circle the 10s digit.

4 3 7

2. Write the amount.

Q  

_____¢

3. Write the number.

卌 卌 卌 卌 ||||

_____

4. Write the missing number.

$$\begin{array}{r} 7 \\ + \square \\ \hline 10 \end{array}$$

5. Write the missing numbers.

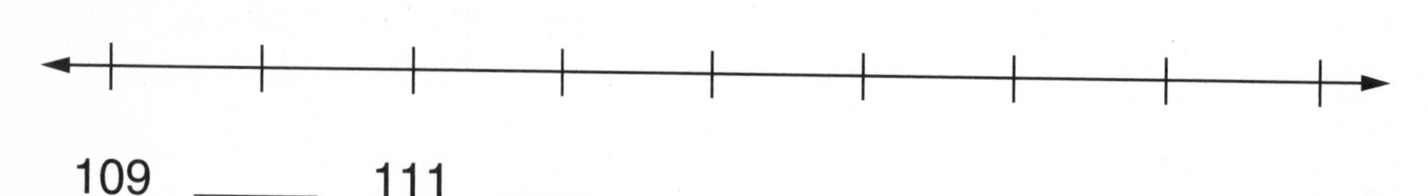

109 _____ 111 _____ _____ _____ _____ _____ _____

6. Write the missing numbers.

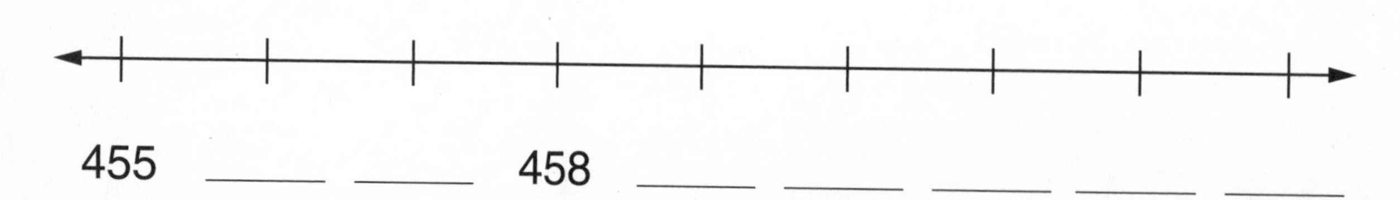

455 _____ _____ 458 _____ _____ _____ _____ _____

Date Time

Number-Grid Puzzles

80

9

7

66

35

2

21

61

Date Time

Number-Grid Hunt

401		403		⏢	□	○	408	⏢	△
○	⏢	□		415		○		419	□
421	○		424		△		⏢		⏢
□	△		⏢	□	○	437		△	440
⏢	442	⏢	○		446		○	⏢	
△		△		455		⏢	□	459	○
	462	○	□	⏢		□	△		470
□	⏢		474	△	476		○	□	⏢
⏢	□	483			□	⏢	488		○
491	○	⏢	△	○		497			△

Draw the shape (⏢, ○, □, or △) for the number.

1. 406 □
2. 422 ______
3. 448 ______
4. 500 ______
5. 431 ______
6. 486 ______
7. 430 ______
8. 479 ______
9. 457 ______
10. 492 ______
11. 493 ______
12. 468 ______

Date Time

Math Boxes 1.9

1. Write the number that is 10 more.

42 ______

57 ______

2. Solve.

______ + 6 = 8

3. Write the amount.

______¢

4. Draw the hands so the clock shows 4:45.

5. Circle the digit in the 100s place.

8 4 9

6. Write these numbers in order. Start with the smallest number.

103 29 86

______ ______ ______

Date Time

Broken Calculator

Example: Show 17. Broken key is (7). Show several ways: 11 + 6 / 20 − 3 / 8 + 8 + 1	**1.** Show 20. Broken key is (2). Show several ways: ______ ______ ______
2. Show 3. Broken key is (3). Show several ways: ______ ______ ______	**3.** Show 22. Broken key is (2). Show several ways: ______ ______ ______
4. Show 12. Broken key is (1). Show several ways: ______ ______ ______	**5.** Make up your own. Show ______. Broken key is ______. Show several ways: ______ ______ ______

Use with Lesson 1.10.

Date Time

Math Boxes 1.10

1. Show four ways to make 25 cents using Ⓠ, Ⓓ, and Ⓝ.

2. Put an X on the digit in the 10s place.

2 0 5

3. Show 28 with tally marks.

____

4. Write the time.

____ : ____

Fill in the missing numbers.

5.

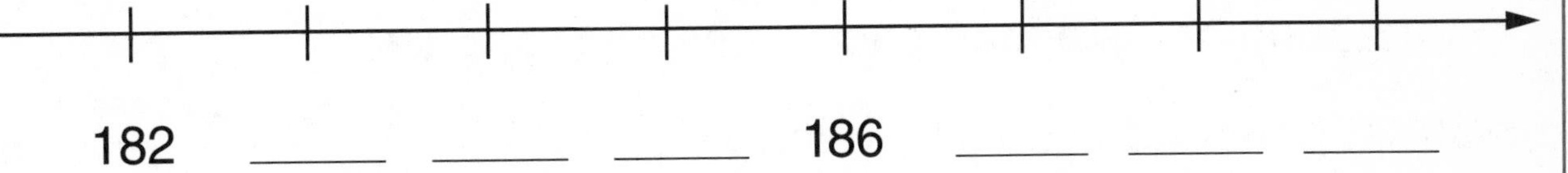

182 ____ ____ ____ 186 ____ ____ ____

6.

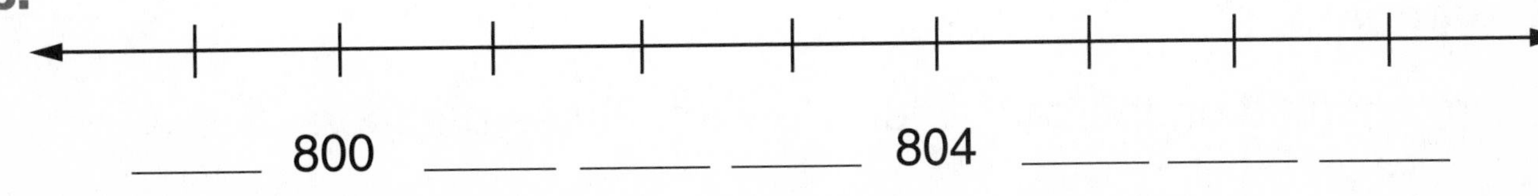

____ 800 ____ ____ ____ 804 ____ ____ ____

Date Time

Counting with a Calculator

Clear the calculator.	(ON/C)
Enter the starting number.	[number]
Tell the calculator to count up or down.	(+) or (−)
Tell the calculator what number to count by.	[number]
Count by pressing the (=) key.	(=)

1. Count by 7s on your calculator. Write the numbers.

   7, ____, ____, ____, ____, ____, ____

   What number is added each time you press (=)? ____

2. Count by 6s on your calculator. Write the numbers.
   Circle the 1s digit in each number.

   6, ____, ____, ____, ____, ____, ____, ____, ____, ____

   What number is added each time you press (=)? ____

   What pattern do you see in the 1s digits? ____________

3. Count by 4s on your calculator. Write the numbers.
   Circle the 1s digit in each number.

   4, ____, ____, ____, ____, ____, ____, ____, ____, ____

   What number is added each time you press (=)? ____

   What pattern do you see in the 1s digits? ____________

Challenge

4. Jim counted on the calculator. He wrote these numbers: 3, 5, 7, 9, 11.

   What keys did he press? ____________________

Date Time

Math Boxes 1.11

1. Fill in the missing numbers.

89	
	100

56	
	67

2. Write the amount.

_____ ¢

3. Write the time.

_____ : _____

4. Put an X on the digit in the 100s place.

9 3 6

5. Write the number that is 10 more.

14 _____

28 _____

103 _____

6. How many?

~~||||~~ ~~||||~~ ~~||||~~ ~~||||~~ ~~||||~~ ~~||||~~ ||||

_____

Date Time

Using <, >, and =

$3 < 5$ 3 is less than 5.	$5 > 3$ 5 is greater than 3.

Write <, >, or =.

1. 61 ______ 26 **2.** 18 ______ 81 **3.** 107 ______ 57

4. 114 ______ 114 **5.** 299 ______ 302 **6.** 801 ______ 688

7. 15 ______ 7 + 8 **8.** 9 + 2 ______ 4 + 5 **9.** 5 + 6 ______ 8 + 4

---

10. Write the total amounts. Then write <, >, or =.

Example: Ⓓ Ⓝ Ⓝ Ⓟ Ⓟ = 22¢ < 26¢ = Ⓠ Ⓟ

a. Ⓝ Ⓝ Ⓓ Ⓟ = ______¢ ______ ______¢ = Ⓠ Ⓝ Ⓓ Ⓝ

b. Ⓝ Ⓓ Ⓟ Ⓠ = ______¢ ______ ______¢ = Ⓝ Ⓓ Ⓓ Ⓝ Ⓟ Ⓠ

c. Ⓓ Ⓓ Ⓠ Ⓓ = ______¢ ______ ______¢ = Ⓓ Ⓝ Ⓟ Ⓓ

Date Time

Math Boxes 1.12

1. Write 5 names for 15.

2. Write the number.

HHH HHH HHH HHH

HHH HHH HHH /

______

3. Solve.

$$\begin{array}{r} 9 \\ + \square \\ \hline 14 \end{array}$$

4. Fill in the blanks.

55, 60, ______, ______,

______, ______, ______

5. Write the amount.

______¢

6. Draw the hands to show 9:00.

Date Time

Math Boxes 1.13

1. Fill in the missing numbers.

92	
	103

2. Use $<$, $>$, or $=$.

$7 + 8$ ______ $9 + 6$

$15 - 9$ ______ 24

$8 + 8$ ______ $10 + 4$

3. Write 5 names for 18.

4. Continue.

397, 398, 399, ______, ______,

______

5. Put a line under the digit in the ones place.

4 7 9

6. Continue. Circle the even numbers.

85, 90, ______, ______,

______, ______, ______

Date Time

Math Boxes 1.14

1. Fill in names for 12.

   ______ + ______ = 12

   12 = ______ + ______

   12 = ______ + ______

   ______ + ______ = 12

2. How much money?

   $______.______

3. Play *Broken Calculator.* Show 11. Broken key is 1. Show 3 ways.

   ____

   ____

   ____

4. Today is

   ______________ ________, ________.
   (month) (day) (year)

   This month has ______ days.

5. Fill in the missing numbers.

	125	
134		

6. Write the number that is 10 more.

   97 ______

   197 ______

   297 ______

Date Time

Addition Number Stories

Write two addition number stories about things in the picture.

For each story:
- Write a label in the unit box.
- Find the answer. Write a number model.

Example: *7 ducks in the water. 5 ducks in the grass. How many ducks in all?*

Unit
ducks

Answer the question: *12 ducks* (unit)

Number model: *7* + *5* = *12*

Story 1: ______________________________

____

Unit

Answer the question: ________ (unit) Number model: ____ + ____ = ____

Story 2: ______________________________

____

Unit

Answer the question: ________ (unit) Number model: ____ + ____ = ____

Date Time

Number-Grid Puzzles

Complete the number-grid puzzles.

11				15					20
							28		
		33							
	52						58		

	332				336				
								349	
						357			
361									
		373					378		

Date Time

Math Boxes 2.1

1. Use $>$, $<$, or $=$. $6 + 7$ ______ $15 - 4$ $5 + 8$ ______ $8 + 5$ $18 - 9$ ______ $5 + 4$	**2.** Continue the count. Then circle the odd numbers. 598, ______, 600, ______, 602, ______
3. Write the number that is 10 less. 75 ______ 90 ______ 83 ______ 106 ______	**4.** Draw hands to show 8:15. 
5. Solve. 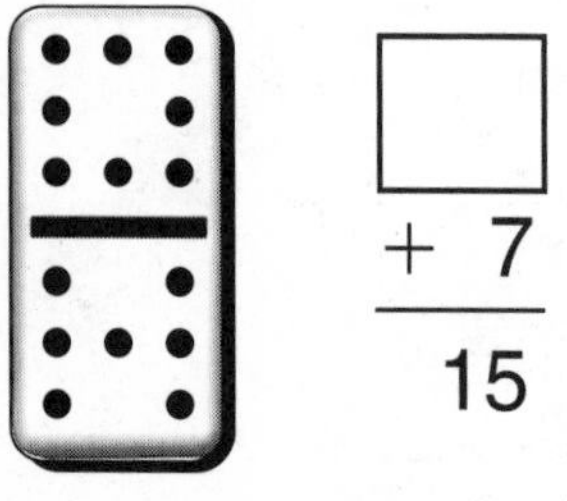 $\square$ $+ 7$ 15	**6.** Show $1.00 three ways. Use Ⓠ, Ⓓ, and Ⓝ.

Date Time

Fact Power Table

0 + 0	0 + 1	0 + 2	0 + 3	0 + 4	0 + 5	0 + 6	0 + 7	0 + 8	0 + 9
1 + 0	1 + 1	1 + 2	1 + 3	1 + 4	1 + 5	1 + 6	1 + 7	1 + 8	1 + 9
2 + 0	2 + 1	2 + 2	2 + 3	2 + 4	2 + 5	2 + 6	2 + 7	2 + 8	2 + 9
3 + 0	3 + 1	3 + 2	3 + 3	3 + 4	3 + 5	3 + 6	3 + 7	3 + 8	3 + 9
4 + 0	4 + 1	4 + 2	4 + 3	4 + 4	4 + 5	4 + 6	4 + 7	4 + 8	4 + 9
5 + 0	5 + 1	5 + 2	5 + 3	5 + 4	5 + 5	5 + 6	5 + 7	5 + 8	5 + 9
6 + 0	6 + 1	6 + 2	6 + 3	6 + 4	6 + 5	6 + 6	6 + 7	6 + 8	6 + 9
7 + 0	7 + 1	7 + 2	7 + 3	7 + 4	7 + 5	7 + 6	7 + 7	7 + 8	7 + 9
8 + 0	8 + 1	8 + 2	8 + 3	8 + 4	8 + 5	8 + 6	8 + 7	8 + 8	8 + 9
9 + 0	9 + 1	9 + 2	9 + 3	9 + 4	9 + 5	9 + 6	9 + 7	9 + 8	9 + 9

Date Time

Distances on a Number Grid

Example: *How many spaces do you move to go from 17 to 23 on the number grid?*

Solution: *Place a marker on 17. You move the marker 6 spaces before landing on 23.*

11	12	13	14	15	16	17	18	19	20
21	22	23	24	25	26	27	28	29	30

How many spaces from:

23 to 28? _____ 15 to 55? _____ 39 to 59? _____

27 to 42? _____ 34 to 26? _____ 54 to 42? _____

15 to 25? _____ 26 to 34? _____

1	2	3	4	5	6	7	8	9	10
11	12	13	14	15	16	17	18	19	20
21	22	23	24	25	26	27	28	29	30
31	32	33	34	35	36	37	38	39	40
41	42	43	44	45	46	47	48	49	50
51	52	53	54	55	56	57	58	59	60

Date Time

Domino-Dot Patterns

Draw the missing dots on the dominoes. Find the total number on both halves.

1. double 2

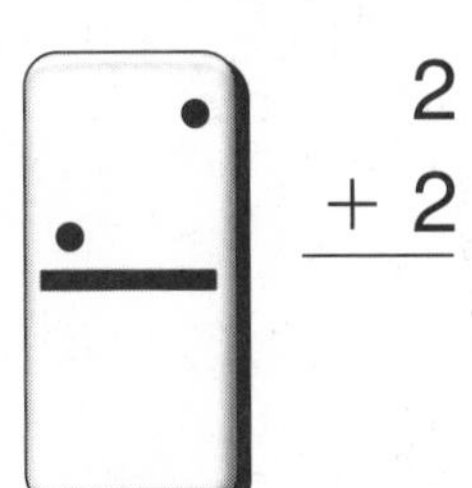

$$\begin{array}{r} 2 \\ +\ 2 \\ \hline \end{array}$$

2. double 3

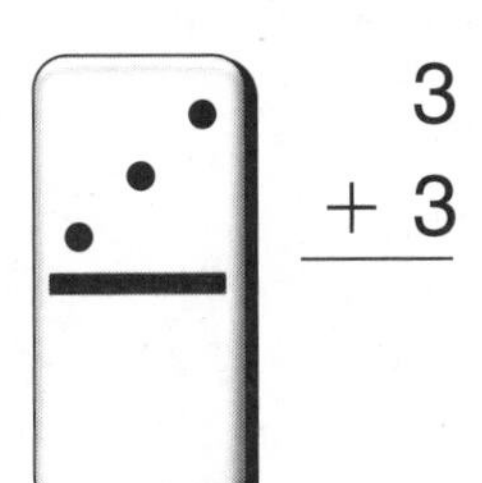

$$\begin{array}{r} 3 \\ +\ 3 \\ \hline \end{array}$$

3. double 4

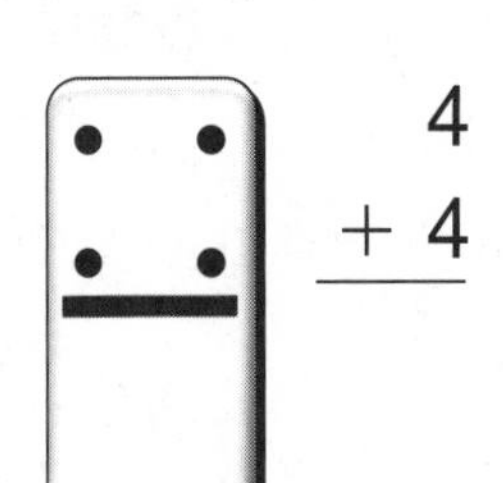

$$\begin{array}{r} 4 \\ +\ 4 \\ \hline \end{array}$$

4. double 5

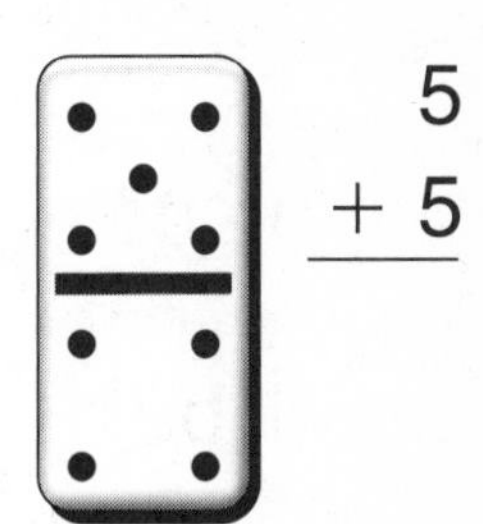

$$\begin{array}{r} 5 \\ +\ 5 \\ \hline \end{array}$$

5. double 6

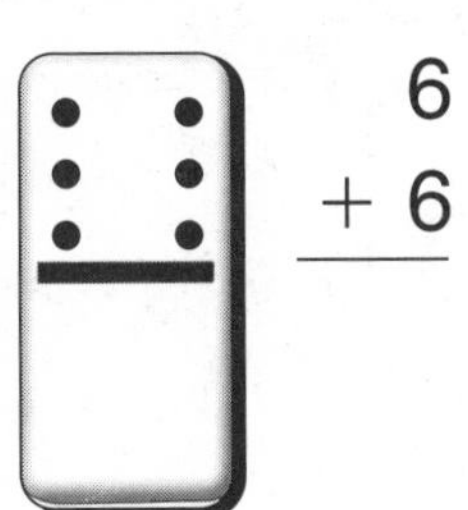

$$\begin{array}{r} 6 \\ +\ 6 \\ \hline \end{array}$$

6. double 7

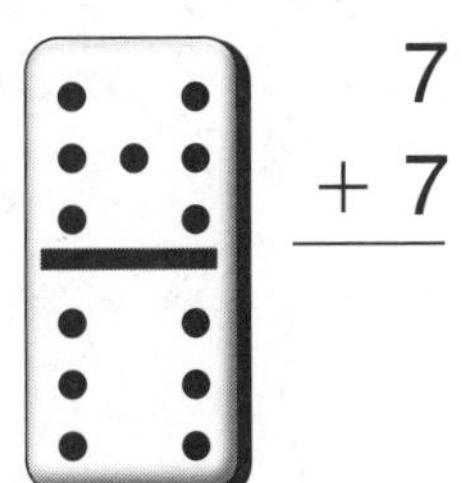

$$\begin{array}{r} 7 \\ +\ 7 \\ \hline \end{array}$$

7. double 8

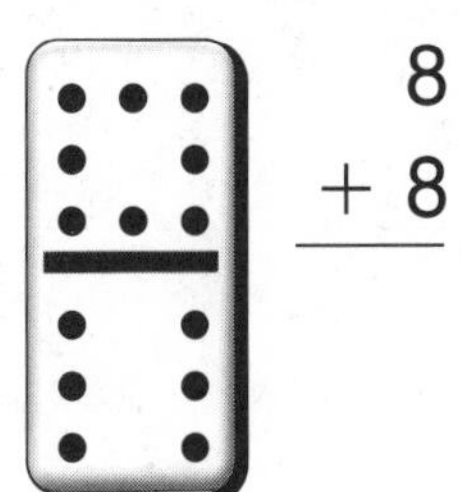

$$\begin{array}{r} 8 \\ +\ 8 \\ \hline \end{array}$$

8. double 9

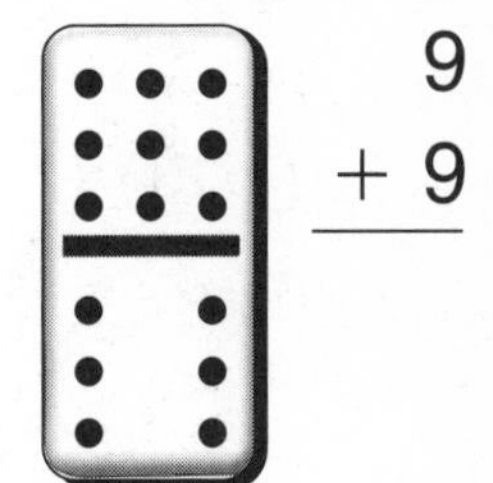

$$\begin{array}{r} 9 \\ +\ 9 \\ \hline \end{array}$$

Find the total number of dots.

9.

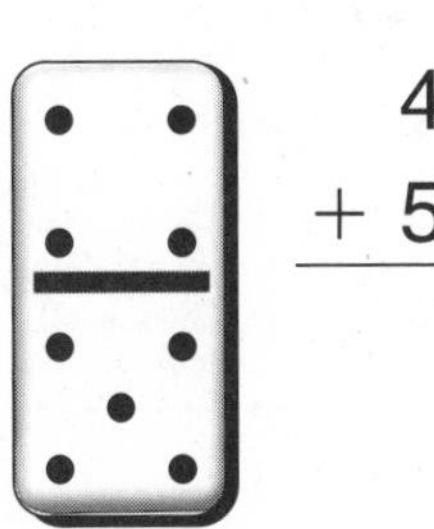

$$\begin{array}{r} 4 \\ +\ 5 \\ \hline \end{array}$$

10.

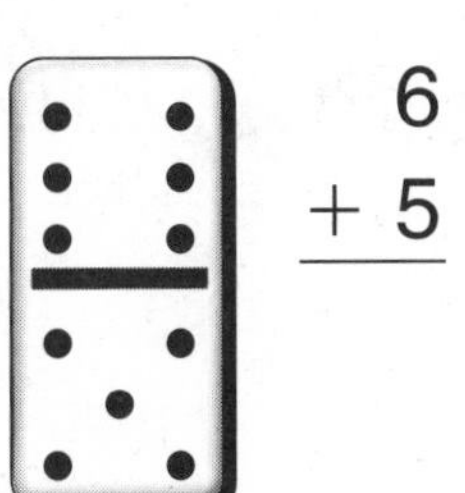

$$\begin{array}{r} 6 \\ +\ 5 \\ \hline \end{array}$$

11.

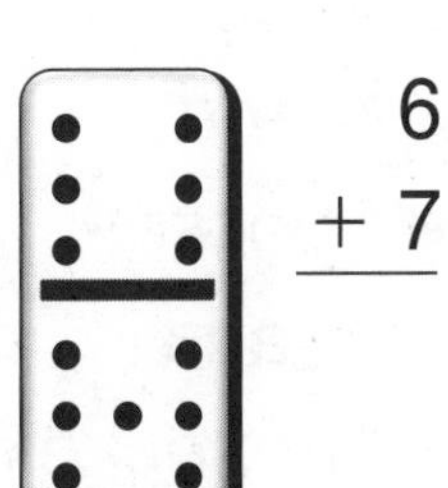

$$\begin{array}{r} 6 \\ +\ 7 \\ \hline \end{array}$$

12.

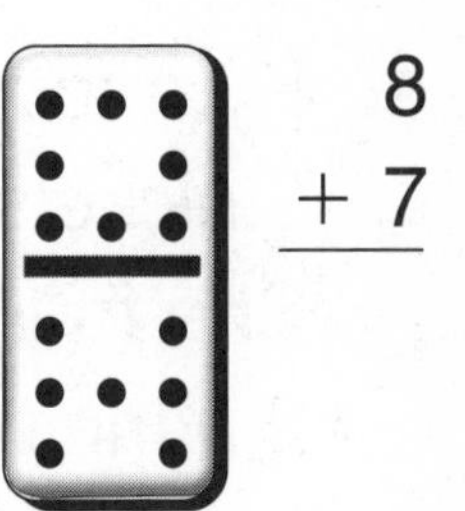

$$\begin{array}{r} 8 \\ +\ 7 \\ \hline \end{array}$$

Date Time

Math Boxes 2.3

1. Today is

________ ______, ______.
(month) (day) (year)

The date 1 week from today

will be ______________.

2. Put a ✓ on the digit in the hundreds place.

2 7 3

3. Count back by 5s.

45, 40, ____, ____, ____,

____, ____, 10, ____

Can you keep going?

0, ____, ____

4. Solve.

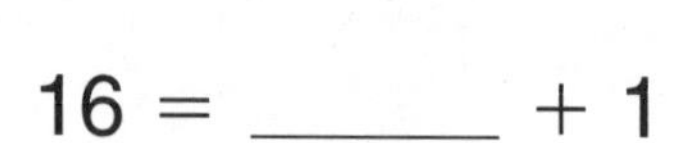

Unit

____ + 9 = 9

16 = ____ + 1

____ = 14 + 0

18 = 1 + ____

5. Kyra found 2 dimes and 3 nickels in her left pocket. She found 1 quarter and 2 pennies in her right pocket. How much money did she find?

____

6. Show tallies for 42.

____

____

Date Time

+9 Facts

Write the sums.

1. $\begin{array}{r} 3 \\ +\ 9 \\ \hline \end{array}$

2. $\begin{array}{r} 7 \\ +\ 9 \\ \hline \end{array}$

3. ____ $= 9 + 5$

4. ____ $= 2 + 9$

5. $\begin{array}{r} 9 \\ +\ 4 \\ \hline \end{array}$

6. $\begin{array}{r} 6 \\ +\ 9 \\ \hline \end{array}$

7. $\begin{array}{r} 9 \\ +\ 8 \\ \hline \end{array}$

8. $1 + 9 =$ ____

9. ____ $= 0 + 9$

10. $\begin{array}{r} 9 \\ +\ 9 \\ \hline \end{array}$

Write the missing numbers.

11. $9 +$ ____ $= 12$

12. $15 = 9 +$ ____

13. ____ $+ 9 = 17$

14. $13 =$ ____ $+ 9$

15. $16 =$ ____ $+ 9$

16. $14 = 9 +$ ____

Riddle Decoder

17. To solve the riddle, first write the sums. Then write the letter for each sum in the boxes.

Riddle: *What kind of dog has ticks?*

$\begin{array}{r} 9 \\ +\ 9 \\ \hline \end{array}$ $\begin{array}{r} 2 \\ +\ 2 \\ \hline \end{array}$ $\begin{array}{r} 8 \\ +\ 8 \\ \hline \end{array}$ $\begin{array}{r} 3 \\ +\ 3 \\ \hline \end{array}$ $\begin{array}{r} 6 \\ +\ 6 \\ \hline \end{array}$ $\begin{array}{r} 4 \\ +\ 4 \\ \hline \end{array}$ $\begin{array}{r} 7 \\ +\ 7 \\ \hline \end{array}$ $\begin{array}{r} 5 \\ +\ 5 \\ \hline \end{array}$

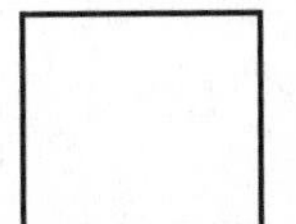 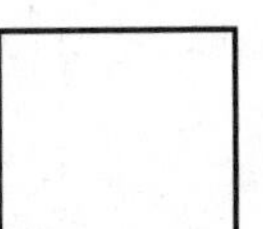 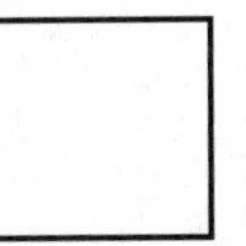  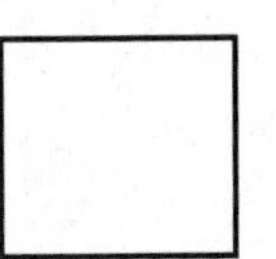 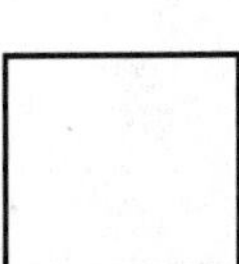  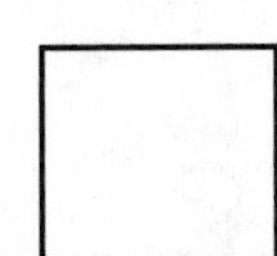

Sum	4	6	8	10	12	14	16	18 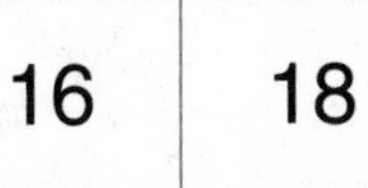
Letter	A	C	D	G 	H 	O	T 	W

Date Time

Math Boxes 2.4

1. Write the sums.

7 + 0 = _____ _____ = 7 + 1

8 + 0 = _____ _____ = 8 + 1

9 + 0 = _____ _____ = 9 + 1

10 + 0 = _____ _____ = 10 + 1

2. Write 4 doubles facts that you know.

____

____

____

____

3. Tonisha had 9 dominoes. She found 8 more. How many does she have now?

_____ dominoes

4. Use a number grid. How many spaces from:

17 to 26? _____

49 to 28? _____

5. Draw hands to show 6:55.

6. Write the sums.

Unit

5 + 7 = _____

6 + 8 = _____

_____ = 9 + 6 6 + 7 = _____

9 + 8 = _____ _____ = 5 + 4

 Use with Lesson 2.4.

Date Time

Addition Facts

If you know a double, you know the *1-more* and the *1-less* sums.

Reminder: If you know that $4 + 4 = 8$,
you know $4 + \mathbf{5} = 9$,
and $4 + \mathbf{3} = 7$.

For Problems 1–10, find the sums.

1. ____ $= 8 + 7$

2. $3 + 4 =$ ____

3. ____ $= 6 + 7$

4. $\begin{array}{r} 6 \\ +\ 5 \\ \hline \end{array}$

5. $\begin{array}{r} 8 \\ +\ 9 \\ \hline \end{array}$

6. $\begin{array}{r} 7 \\ +\ 5 \\ \hline \end{array}$

7. $\begin{array}{r} 7 \\ +\ 9 \\ \hline \end{array}$

8. $5 + 8 =$ ____

9. ____ $= 6 + 9$

10. $8 + 6 =$ ____

Try these.

11. $8 + 8 = 16$

$8 + \mathbf{9} =$ ____

$8 + \mathbf{7} =$ ____

12. $12 + 12 = 24$

$12 + \mathbf{13} =$ ____

$12 + \mathbf{11} =$ ____

13. $15 + 15 = 30$

$\mathbf{16} + 15 =$ ____

$\mathbf{14} + 15 =$ ____

14. Can you figure these out?

$\mathbf{14} + 12 =$ ____

$15 + \mathbf{13} =$ ____

Date Time

Math Boxes 2.5

1. Write the turnaround for each fact.

$7 + 9 = 16$ 9 + 7 = 16

$15 = 6 + 9$ ______

$17 = 9 + 8$ ______

2. Julie had 10 crayons. Rosa gave her 8 more crayons. How many crayons in all?

______ crayons

3. Write the numbers in order. Write the smallest number first.

56, 32, 75, 21

______, ______, ______, ______

4. Fill in the missing numbers.

<table>
<tr><td>196</td><td></td><td style="border:none"></td></tr>
<tr><td style="border:none"></td><td>207</td><td style="border:none"></td></tr>
<tr><td style="border:none"></td><td></td><td></td></tr>
<tr><td style="border:none"></td><td></td><td></td></tr>
</table>

5. Solve.

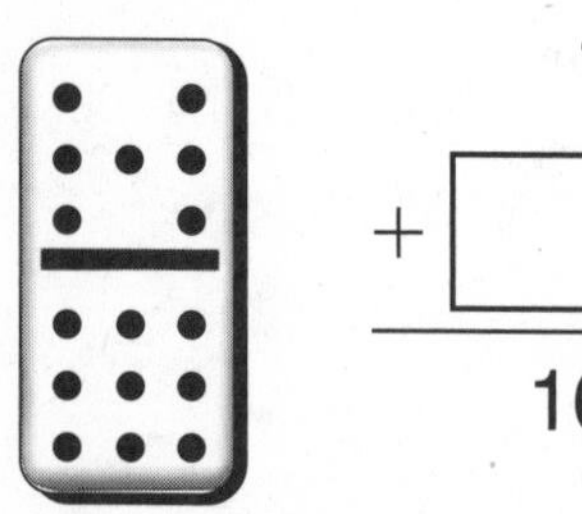

$$\begin{array}{r} 7 \\ + \square \\ \hline 16 \end{array}$$

6. Write the number that is 10 more.

104 ______

76 ______

80 ______

47 ______

Date Time

Domino Facts

For Problems 1–7, write 2 addition facts and 2 subtraction facts for each domino.

1.

$4 + 2 = 6$ $2 + 4 = 6$ $6 - 2 = 4$ $6 - 4 = 2$

2.

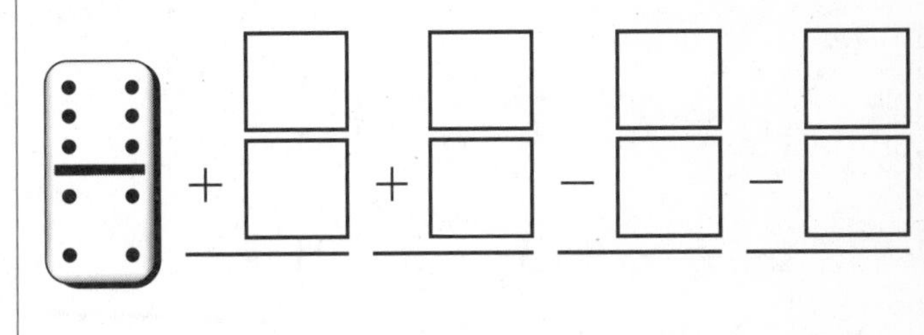

3.

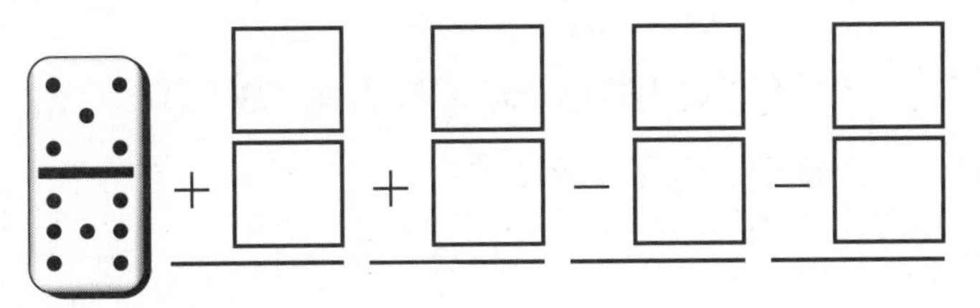

4.

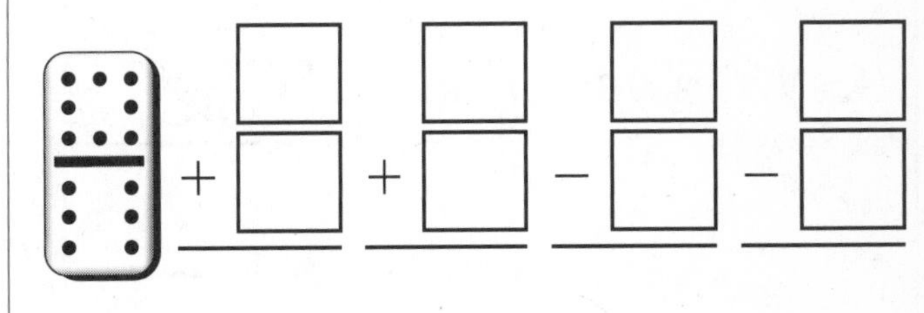

5.

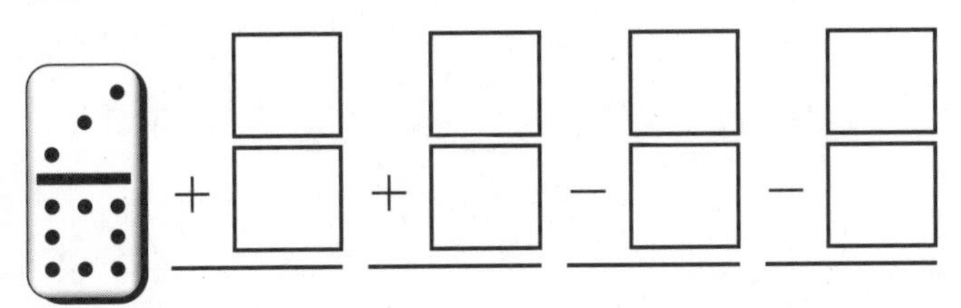

6.

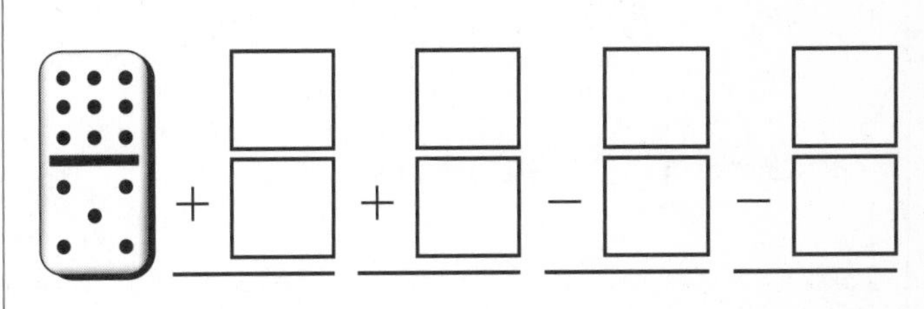

7.

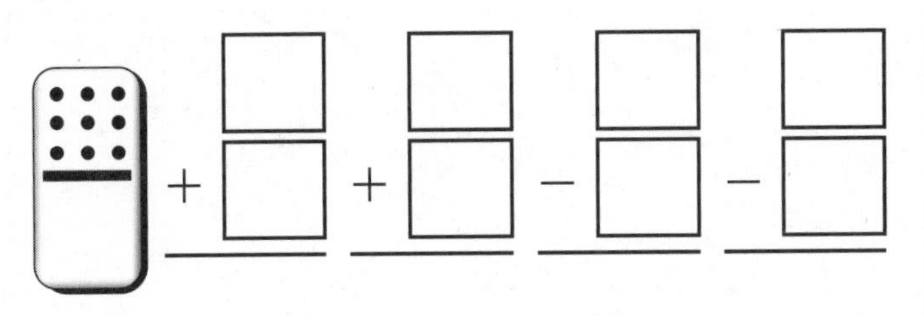

8. Write one addition fact and one subtraction fact.

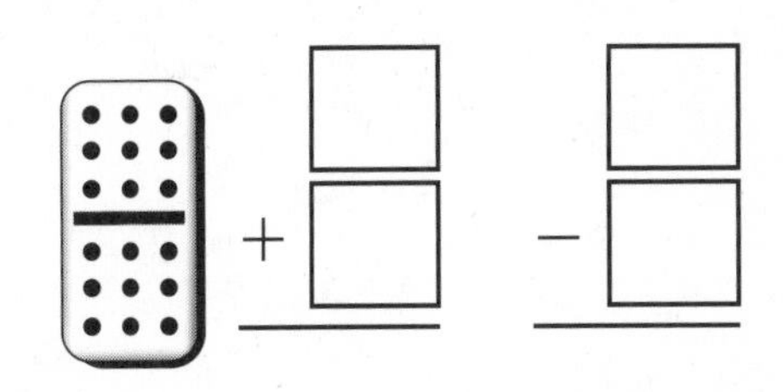

Date Time

Math Boxes 2.6

1. Solve.

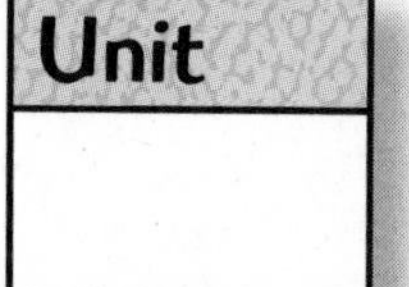

____ $= 7 + 7$ $9 + 9 =$ ____

____ $= 7 + 6$ $9 + 8 =$ ____

____ $= 7 + 8$ $9 + 10 =$ ____

2. Write 5 facts and the turnaround for each.

____ ____

____ ____

____ ____

____ ____

____ ____

3. Write *yes* or *no*.

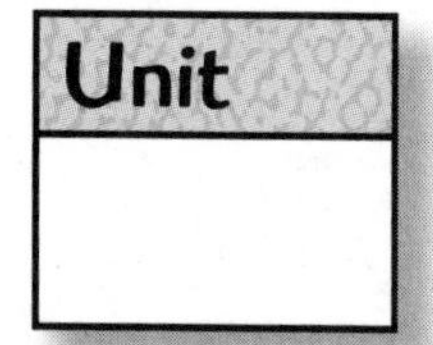

$9 + 3 = 3 + 9$ ____

$12 - 5 = 5 + 12$ ____

$8 + 12 = 12 + 8$ ____

4. Use a number grid. How many spaces from:

9 to 19? ____

47 to 38? ____

53 to 42? ____

5. Continue.

362, ____, ____, ____,

366, ____, ____, ____,

____, ____, ____

6. 219

Write the number that is 1 more.

____

Date Time

Using a Pan Balance and a Spring Scale

Weighing Things with a Pan Balance

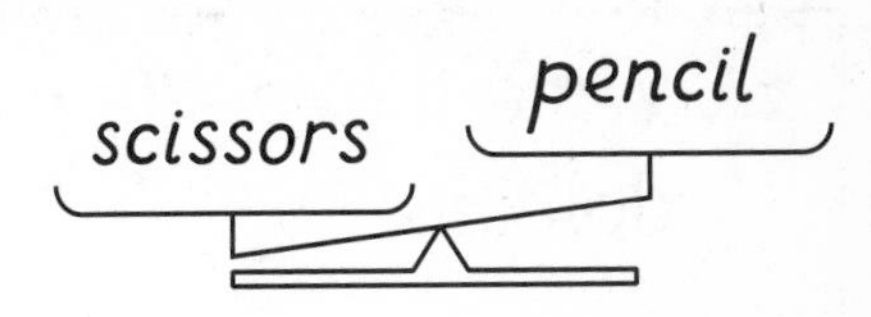

1. Pick two objects. Which feels heavier?

   ____

2. Put one of these objects in the left pan of the pan balance.
3. Put the other object in the right pan.
4. Show what happened on one of the pan-balance pictures.
   - Write the names of the objects on the pan-balance picture.
   - Draw a circle around the pan with the heavier object.
5. Repeat with other pairs of objects.

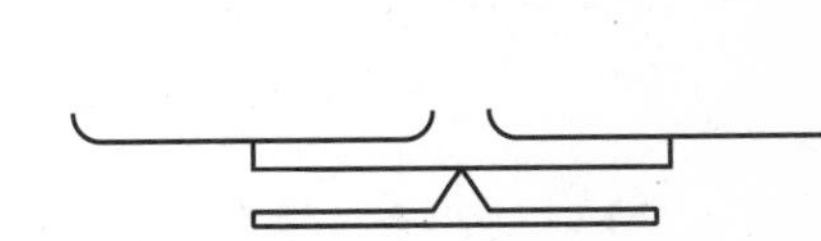

Weighing Things with a Spring Scale

1. Which is heavier: 1 ounce or 1 pound? ____________
2. How many ounces are in 1 pound? ____________
3. Put objects in the plastic bag on the spring scale.
4. Weigh them. Try to get a total weight of about 1 pound.
5. List the objects in the bag that weigh a total of about 1 pound.

______________ ______________ ______________

______________ ______________ ______________

Date Time

Math Boxes 2.7

1. Write the fact family for the domino.

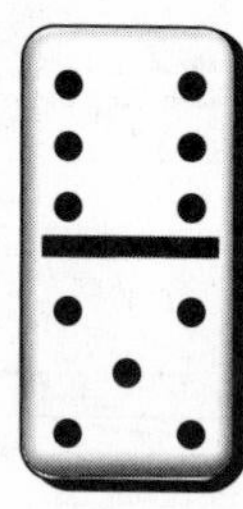

11 = ____ + ____

____ + ____ = 11

11 − ____ = ____

11 − ____ = ____

2. Write the missing number.

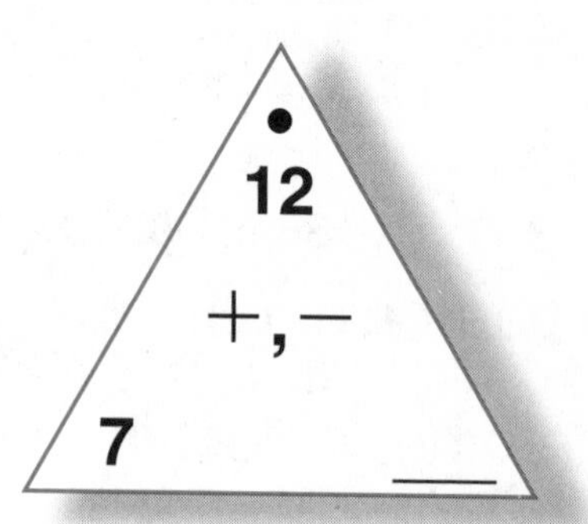

3. Solve.

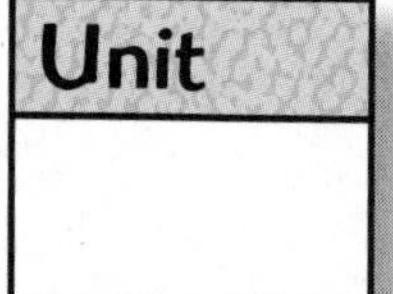

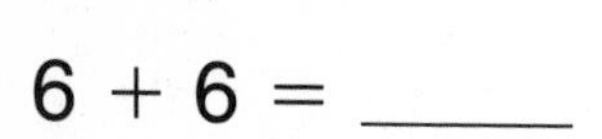

6 + 6 = ____

16 + 6 = ____

16 + 7 = ____

16 + 8 = ____

4. Lee scored 11 points. Oliver scored 18 points.

Who scored more points?

____

How many more?

____ more

5. Write an addition story.

____

____

____

____

6. Solve. Circle the odd sums.

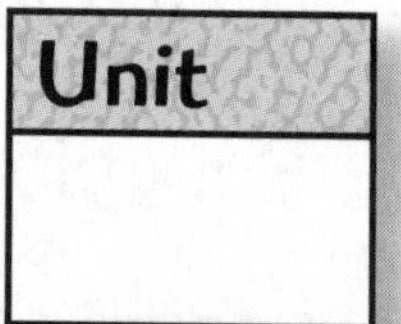

____ = 15 + 9

____ = 22 + 9

36 = ____ + 9

Date Time

Math Boxes 2.8

1. Match the items to the weights.

10 oranges	about 2 ounces
3 pencils	about 55 pounds
1 second grader	about 2 pounds

2. Play *Broken Calculator.* Show 17. Broken key is 2. Show 3 ways.

____

____

____

3. 810

Write the number that is 1 more.

_____

4. Write the sums.

Unit

$10 + 10 =$ _____

$10 + 11 =$ _____

$10 + 12 =$ _____

$10 + 13 =$ _____

5. Nate has 17 toy cars. Raul has 2 fewer toy cars than Nate. How many cars does Raul have?

_____ cars

Write a number model.

____

6. Solve.

Unit

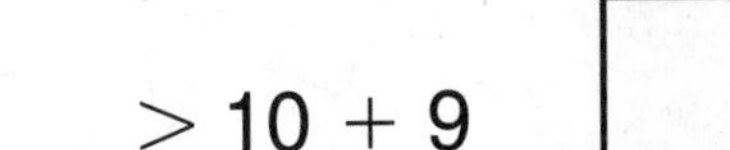
_____ $> 10 + 9$

_____ $< 9 + 8$

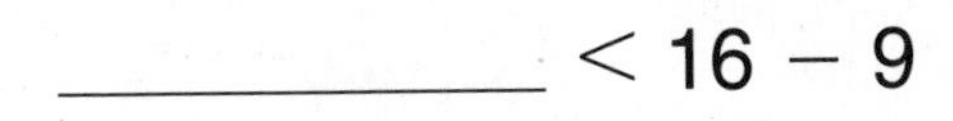
_________ $< 16 - 9$

_________ $= 19 - 8$

Date Time

Pan-Balance Problems

Reminder: There are 16 ounces in 1 pound.

Some food items and their weights are shown below.

- Pretend you will put one or more items in each pan.
- Pick items that would make the balances tilt the way they are shown on journal page 39.
- Write the name of each item in the pan you put it in.
- Write the weight of each item below the pan you put it in.

Try to use a variety of food items.

Salad Dressing
1 ounce

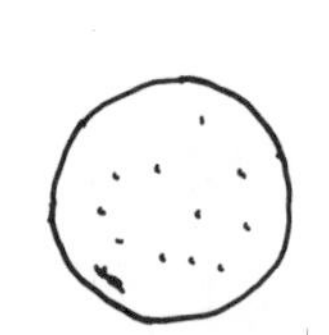

Orange
8 ounces

Walnuts
3 ounces

Eggplant
15 ounces

Gummy Worms
4 ounces

Salt
1 pound

Lemon
6 ounces

Flour
2 pounds

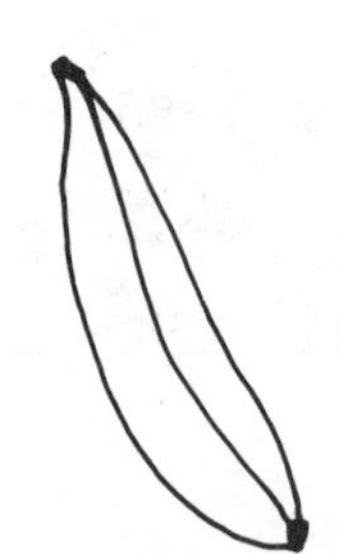

Banana
6 ounces

Potatoes
5 pounds

Date Time

Pan-Balance Problems (cont.)

Example

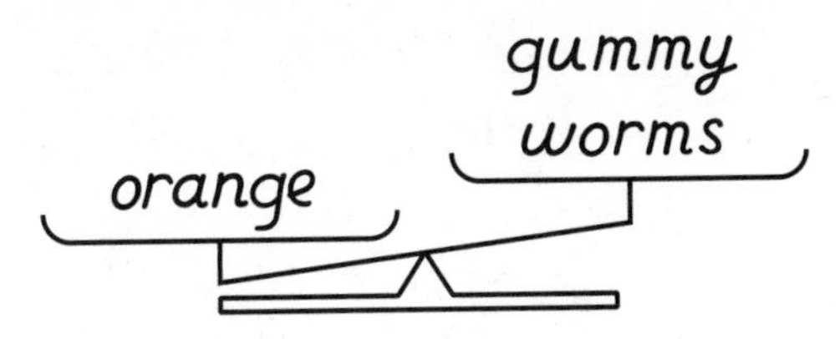

Weight: 8 ounces Weight: 4 ounces

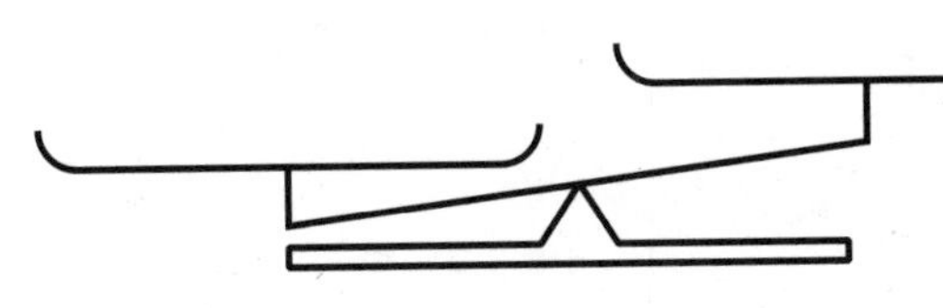

1. Weight: ________ Weight: ________

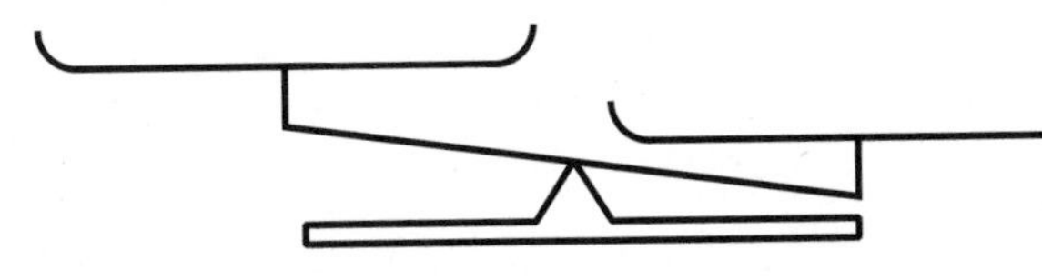

2. Weight: ________ Weight: ________

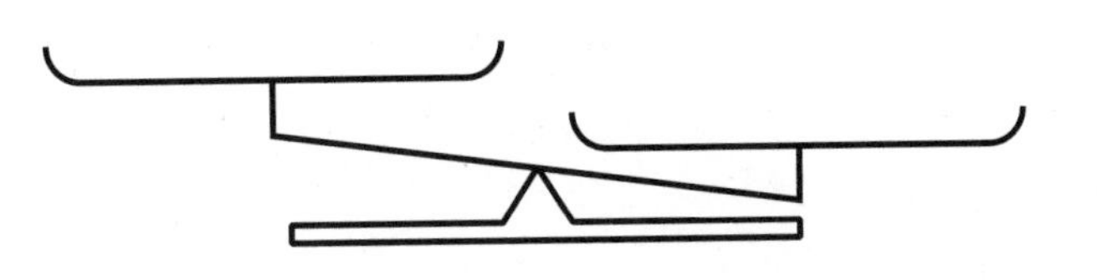

3. Weight: ________ Weight: ________

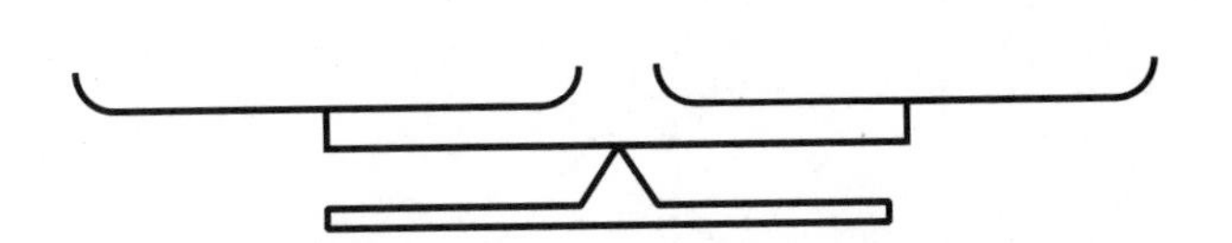

4. Weight: ________ Weight: ________

Date Time

Name-Collection Boxes

1. Write 10 names in the 12-box.

12

2. Circle the names that DO NOT belong in the 9-box.

9

12 − 3
8 + 0
9 − 0
5 + 4 + 1
19 − 10
𝍸 ///
15 − 7
x x x
x x x
x x x
1 less than 10
3 + 3 + 3
nine

3. Three names DO NOT belong in this box. Circle them. Write the name of the box on the tag.

9 + 3
12 − 8
3 + 3
𝍸 //
x x x
x x x
5 + 3 − 2
half a dozen
10 − 4

4. Make up a name-collection box of your own.

Date Time

Subtraction Number Stories

Solve each problem.

1. Ross has $11. He buys a book for $6. How much money does he have left?

   $________

2. Martin has 7 markers. Jason has 4 markers. How many more markers does Martin have than Jason?

   ________ markers

3. There are 11 girls on Tina's softball team. There are 13 girls on Lisa's team. How many more girls are on Lisa's team than on Tina's?

   ________ girls

4. Lily has 10 flowers. She gives 4 flowers to her sister. How many flowers does she have left?

   ________ flowers

5. Emma has 8 chocolate cookies and 5 vanilla cookies. How many more chocolate cookies does she have than vanilla cookies?

   ________ chocolate cookies

6. Make up and solve your own subtraction story.

   ____

   ____

Date Time

Math Boxes 2.9

1. Write the fact family.

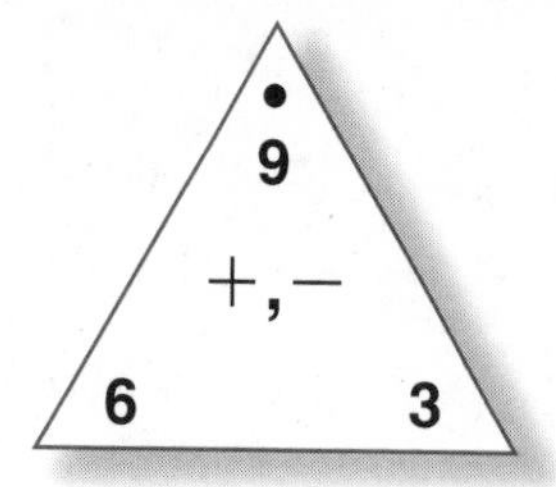

____ = ____ + ____

____ = ____ + ____

____ = ____ − ____

____ = ____ − ____

2. Fill in the missing numbers.

	108
117	

3. Count back by 10s.

220, 210, ____, ____, ____,

____, ____, ____, ____,

____, ____, ____, ____,

____, ____, ____

4. Solve.

Unit

50 + 1 = ____

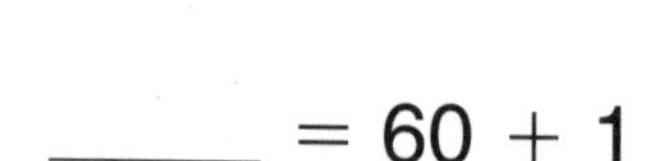

____ = 60 + 1

70 + 1 = ____

____ = 80 + 1

5. Use a number grid. How many spaces from:

99 to 69? ____

74 to 95? ____

53 to 80? ____

6. Shira has 24 crayons. Jasmine has 18 crayons. How many more crayons does Shira have?

____ crayons

Write a number model.

____

Date Time

Frames-and-Arrows Problems

1. Fill in the empty frames.

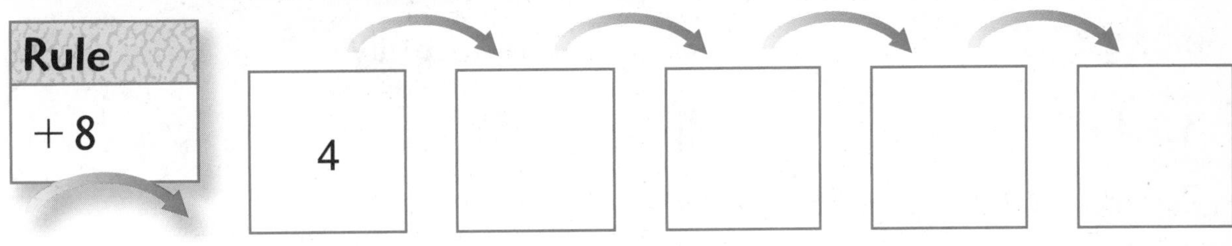

2. Fill in the empty frames.

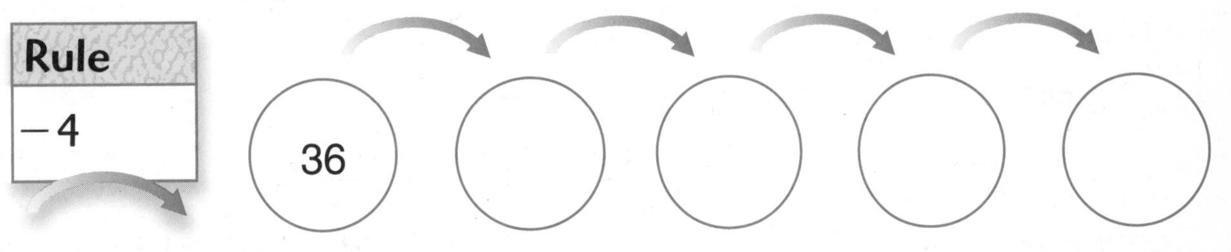

3. Fill in the empty frames.

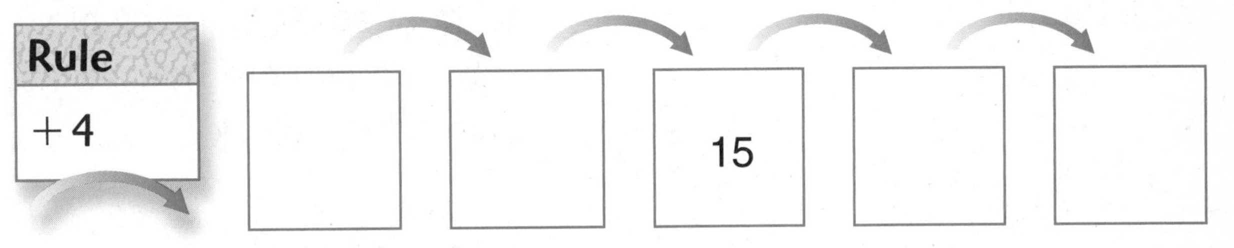

4. Fill in the arrow rule.

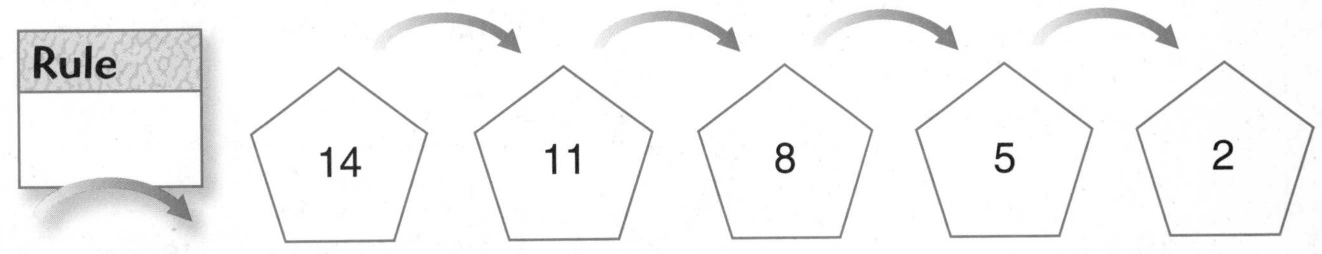

5. Fill in the arrow rule and the empty frames.

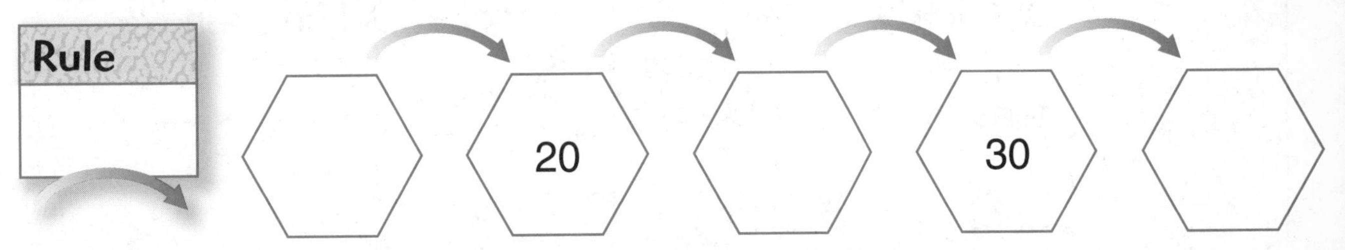

Date Time

Math Boxes 2.10

1. Cross out names that do not belong.

12

9 + 3 10 + 2 18 − 6

卌 卌 || Ⓓ Ⓟ Ⓟ

6 + 5 4 + 4 + 1

1 dozen 3 + 9

2. Write the fact family.

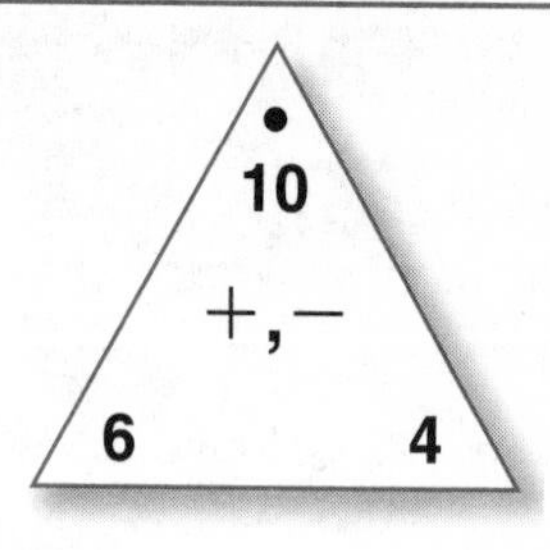

___ − ___ = ___

___ − ___ = ___

___ + ___ = ___

___ + ___ = ___

3. Continue.

230, 235, 240, ____, ____,

____, ____, ____, ____,

____, ____, ____, ____,

____, ____, ____

4. There are 10 houses on Jerry's block. There are 15 houses on Nancy's block. How many more houses are on Nancy's block?

____ houses

Write a number model.

____

5. 349 has:

____ hundreds

____ tens

____ ones

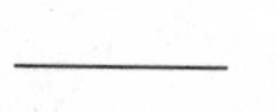

6. A pack of gum costs 25¢. Sean bought 3 packs. How much money did he spend?

____

Date Time

"What's My Rule?"

In Problems 1–4, follow the rule. Fill in the missing numbers.

1.

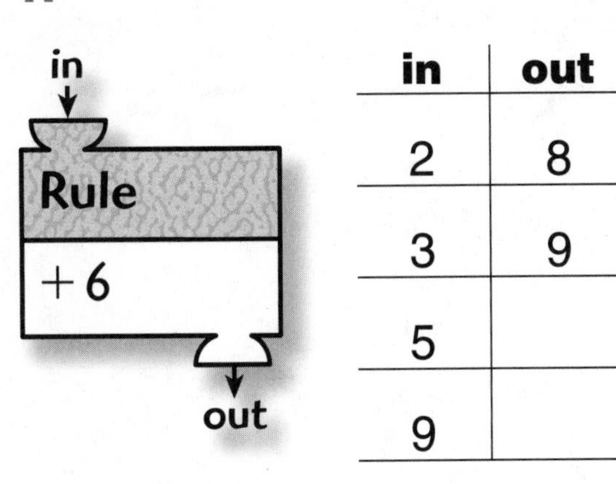

in	out
2	8
3	9
5	
9	

2.

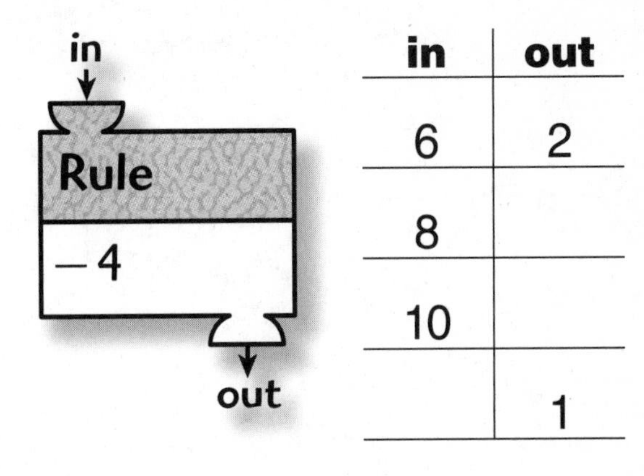

in	out
6	2
8	
10	
	1

3.

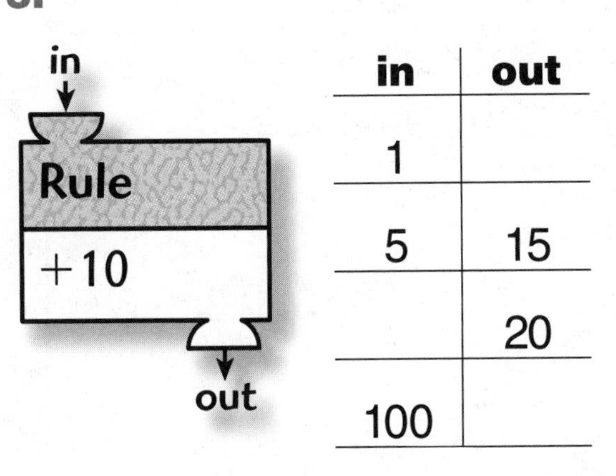

in	out
1	
5	15
	20
100	

4.

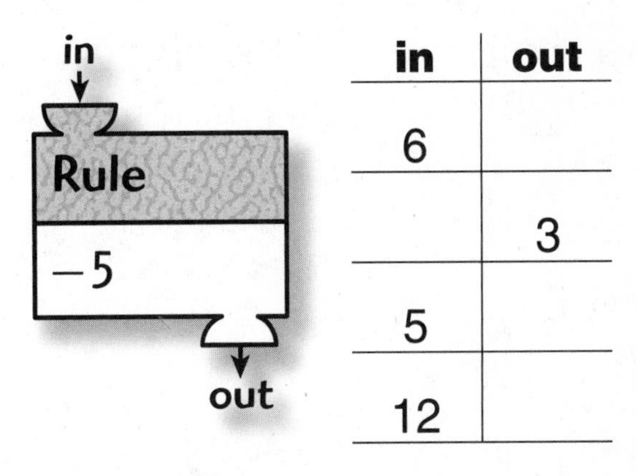

in	out
6	
	3
5	
12	

What is the rule? Write it in the box. Then fill in any missing numbers.

5.

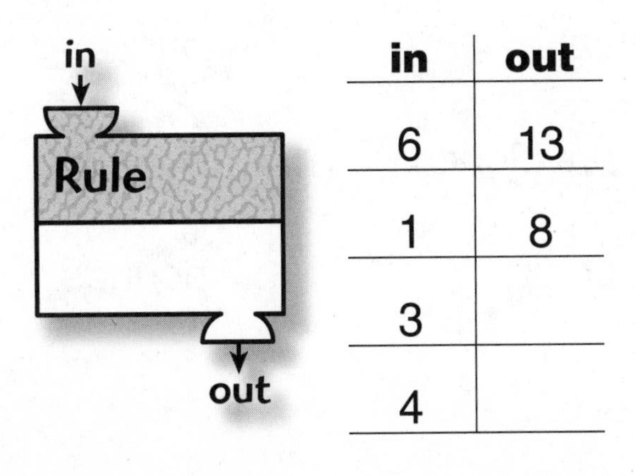

in	out
6	13
1	8
3	
4	

6.

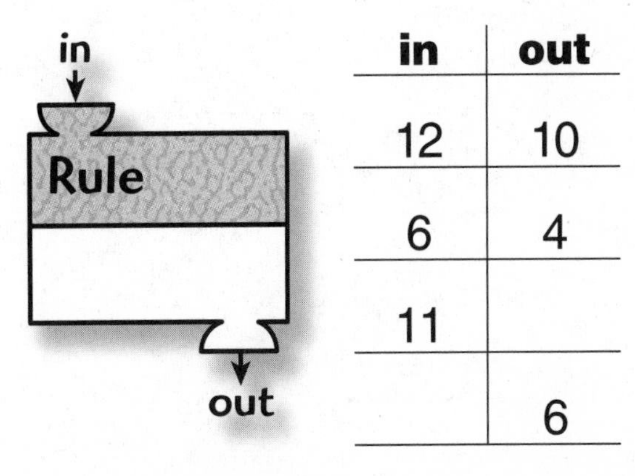

in	out
12	10
6	4
11	
	6

Date Time

Math Boxes 2.11

1. Solve.

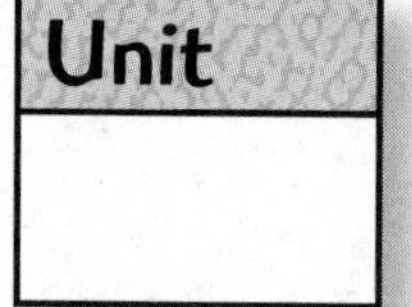

$4 + 3 =$ ____

$10 - 7 =$ ____

$\begin{array}{r} 5 \\ +\ 4 \\ \hline \end{array}$ $\begin{array}{r} 8 \\ -\ 3 \\ \hline \end{array}$

2. Fill in the frames. Follow the arrow rule.

Rule: +3

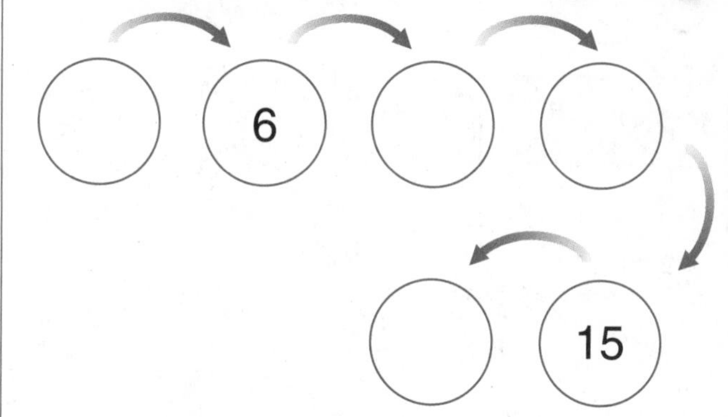

3. Draw a rectangle around the digit in the tens place.

3 4 9

4 0 6

4. Fill in the tag on the name-collection box. Add 3 more names.

$18 - 9$ $3 + 3 + 3$

5. Complete the Fact Triangle and the fact family.

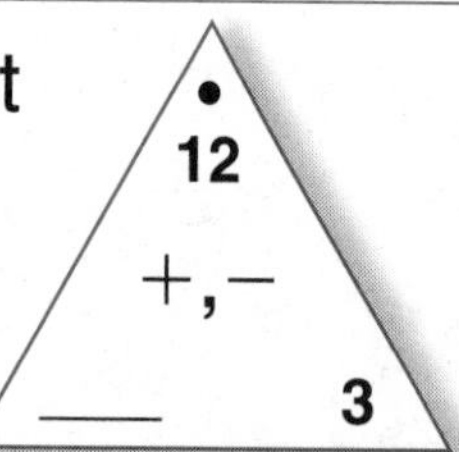

____ = ____ + ____

____ + ____ = ____

____ − ____ = ____

____ = ____ − ____

6. Match the items to the weights.

1 cat	1 ounce
3 envelopes	1 pound
1 book	7 pounds

Math Boxes 2.12

1. Fill in the frames. Follow the arrow rule.

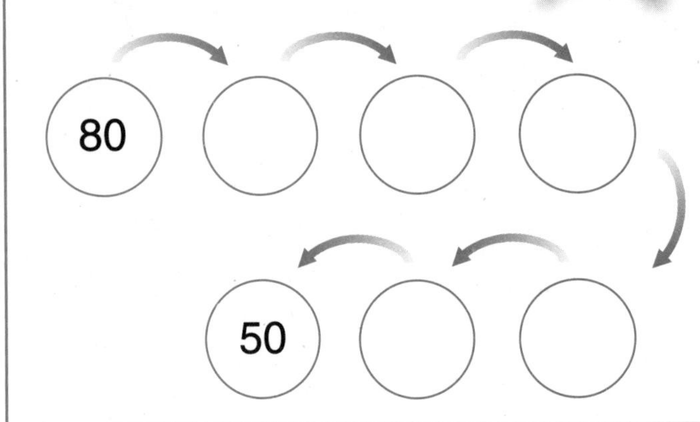

2. Fill in the missing numbers.

144		
	155	

3. Write the fact family.

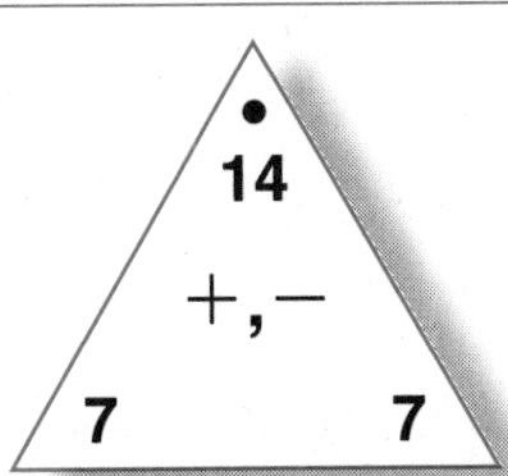

___ − ___ = ___

___ + ___ = ___

4. Put an X on the digit in the tens place.

4 5 6

3 0 9

5. Write these numbers in order. Begin with the smallest number.

133, 146, 129, 151

_____, _____, _____, _____

6. What time is it?

___ : ___

What time was it 1 hour ago?

___ : ___

Date Time

Subtract 9 or 8

1. Subtract. Use the -9 and -8 shortcuts.

a. $13 - 9 =$ ____ b. $14 - 8 =$ ____ c. $16 - 9 =$ ____

d. ____ $= 12 - 8$ e. ____ $= 17 - 9$ f. $12 - 9 =$ ____

g. ____ $= 13 - 8$ h. $11 - 9 =$ ____ i. ____ $= 15 - 8$

j. $\begin{array}{r} 15 \\ -\ 9 \\ \hline \end{array}$ k. $\begin{array}{r} 17 \\ -\ 8 \\ \hline \end{array}$ l. $\begin{array}{r} 11 \\ -\ 8 \\ \hline \end{array}$

2. Find the differences.

> ***Reminder:*** To find $37 - 9$, think $37 - 10 + 1$.
>
> To find $37 - 8$, think $37 - 10 + 2$.

a. $43 - 9 =$ ____ b. $56 - 8 =$ ____ c. $65 - 9 =$ ____

d. $37 - 8 =$ ____ e. $45 - 9 =$ ____ f. $53 - 8 =$ ____

3. Solve.

a. $4 =$ ____ $- 9$ b. $3 =$ ____ $- 8$

c. $7 =$ ____ $- 9$ d. $6 =$ ____ $- 8$

Date Time

Math Boxes 2.13

1. Write the rule in the box. Fill in the missing numbers.

Rule

in	out
6	4
8	6
10	8
2	
12	

2. Solve.

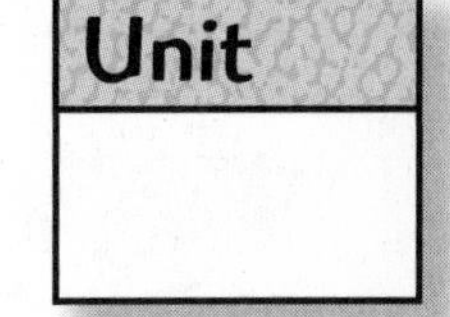

_____ = 13 − 9

14 − 6 = _____

$$\begin{array}{r} 15 \\ -\ 9 \\ \hline \end{array} \qquad \begin{array}{r} 12 \\ -\ 7 \\ \hline \end{array}$$

3. Aisha read 8 books this summer. Pete read 4 more books than Aisha. How many books did Pete read?

_____ books

Write a number model.

____

4. Write 6 names for 13.

13

5. Write the fact family for this domino.

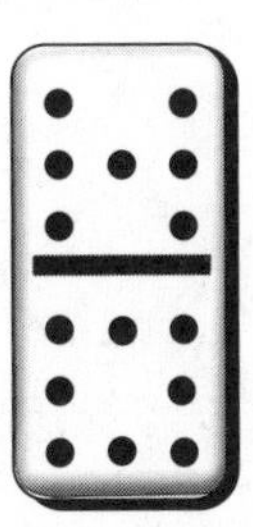

_____ + _____ = _____

_____ + _____ = _____

_____ − _____ = _____

_____ − _____ = _____

6. Write the arrow rule. Fill in the missing frames.

63 → 73 → 83 → ___ → ___ → ___ → ___

Date Time

Math Boxes 2.14

1. How much money?

$____.____

2. Solve.

Unit

3 + 6 = ____

____ = 7 − 5

$$\begin{array}{r} 8 \\ +\ 2 \\ \hline \end{array} \qquad \begin{array}{r} 6 \\ -\ 5 \\ \hline \end{array}$$

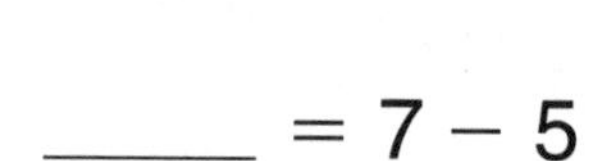

3. Write the rule. Fill in the missing numbers.

Rule

in	out
11	7
9	5
14	

4. Jill earned $18. Kyle earned $9. How much more money did Jill earn?

$______

Write a number model.

____

5. What time is it?

____ : ____

What time will it be in 30 minutes?

____ : ____

6. Fill in the missing frames.

Rule
+2

Date Time

Place Value

Write the number for each group of base-10 blocks.

1. 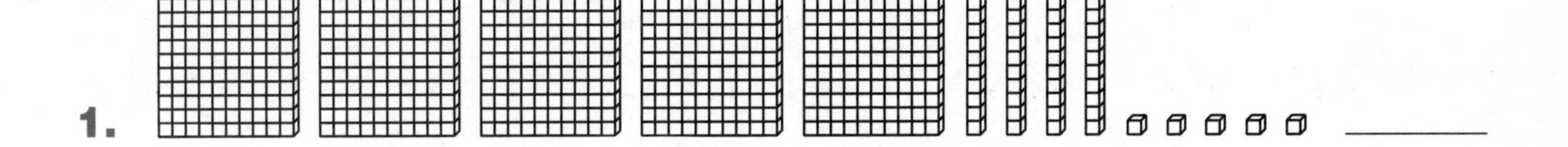 ______

2. 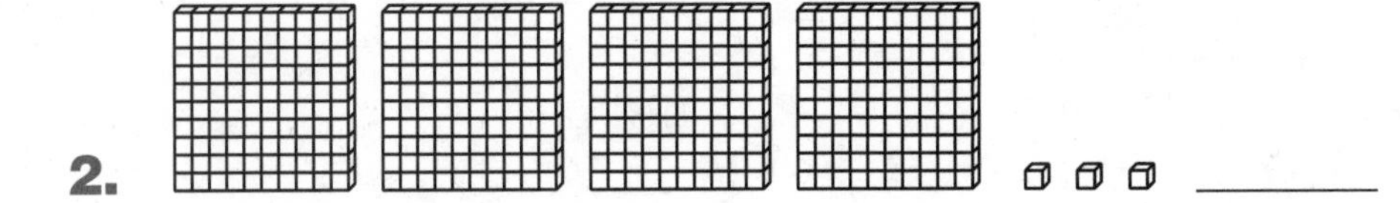 ______

3. Write a number with ...

5 in the ones place

3 in the hundreds place

0 in the tens place ______

4. 960

How many hundreds? ______

How many tens? ______

How many ones? ______

5. Amelia wrote 24 to describe the number shown by these base-10 blocks:

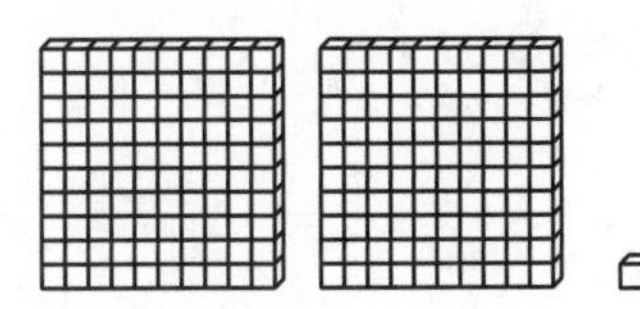

Is Amelia right? Explain your answer.

____

____

____

____

Date Time

Magic Squares

1. Add the numbers in each row. Add the numbers in each column. Add the numbers on each diagonal.

   Are the sums all the same? ____________

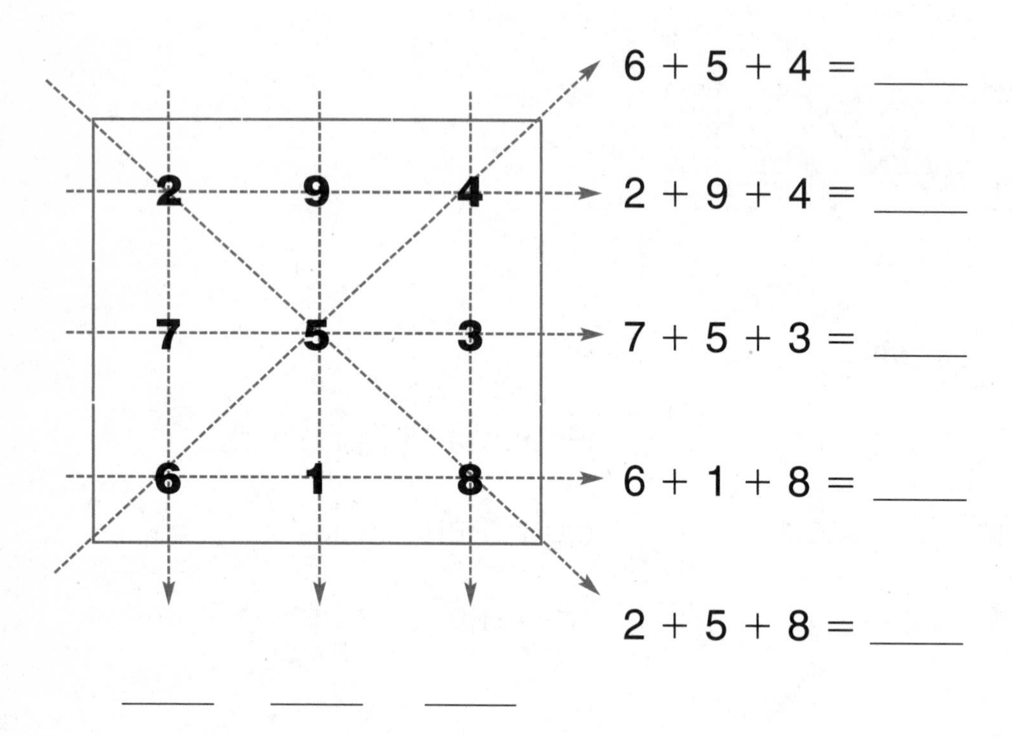

2. The sum of each row, column, and diagonal must be 15. Find the missing numbers. Write them in the blank boxes.

	7	
9		1
4		8

8		6
3		

Date Time

Math Boxes 3.1

1. Write the label and add 3 more names.

20 − 4 ~~||||~~ ~~||||~~ ~~||||~~ |

2. Complete the fact family. Fill in the missing domino dots.

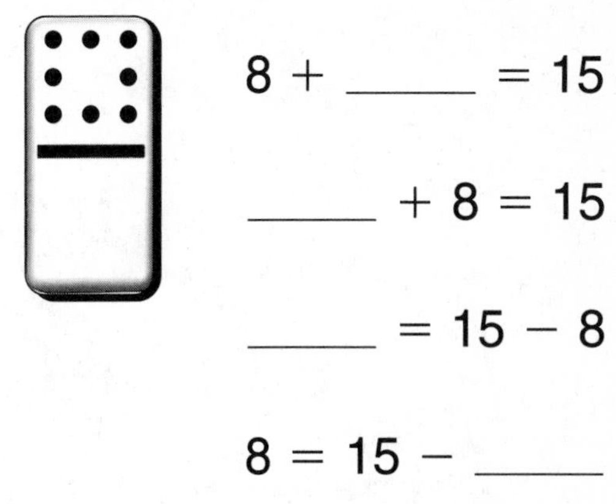

8 + _____ = 15

_____ + 8 = 15

_____ = 15 − 8

8 = 15 − _____

3. Solve.

Unit

5 = _____ − 3

7 = _____ − 3

6 = _____ − 3

4. Jim has 4 pennies. Teri has 3 nickels. Gina has 2 dimes. How much money in all?

_____

5. Fill in the missing rule and numbers.

in	out
48	38
37	27
73	
	104

6. Circle the number models that are true.

$9 + 7 = 7 + 9$

$8 - 5 = 5 - 8$

$6 + 5 = 5 + 6$

Date Time

Fruit and Vegetables Stand Poster

Date Time

Buying Fruit and Vegetables

Select the fruit and vegetables from journal page 54 that you would like to buy. Write the name of each item.

Then draw the coins you could use to pay for each item. Write Ⓟ, Ⓝ, Ⓓ, or Ⓠ.

For Problems 3 and 4, write the total amount of money that you would spend.

I bought (Write the name.)	**I paid (Draw coins.)**	**I paid (Draw coins another way.)**
Example one *orange*	Ⓓ Ⓝ Ⓟ Ⓟ Ⓟ	Ⓝ Ⓝ Ⓟ Ⓟ Ⓟ Ⓟ Ⓟ Ⓟ Ⓟ Ⓟ
1. one ____________		
2. one ____________		
3. one ____________ and one ____________		Total: _____
4. one ____________, one ____________, and one ____________		Total: _____

The *Digit Game*

Materials ❑ 4 cards each of numbers 0–9
(from the Everything Math Deck, if available)

Players 2

Directions

1. Shuffle the deck. Place it facedown between the players.
2. Each player draws 2 cards from the deck and uses them to make the largest number possible.
3. The player who makes the larger number takes all of the cards.
4. The game is over when all of the cards have been used.
5. The player with more cards wins.

Other Ways to Play

A. Players draw 3 cards instead of 2 cards each time.
Each player makes the largest 3-digit number possible.

B. Players try to make the smallest number possible each time.
The person who makes the greatest number takes all of the cards.
The player with fewer cards at the end wins.

Date Time

Math Boxes 3.2

1. Jerry scored 9 points. Lisa scored 7 points. How many points did they score in all?

_____ points

Write a number model.

____

2. 132 has …

_____ hundreds

_____ tens

_____ ones

3. How many in all?

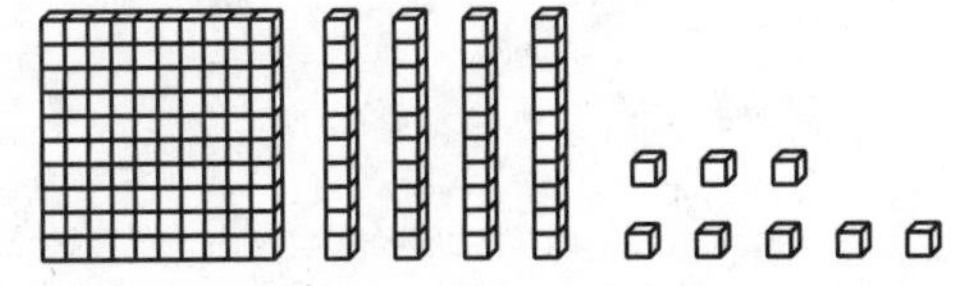

_____

4. How much money?

_____

5. Fill in the frames.

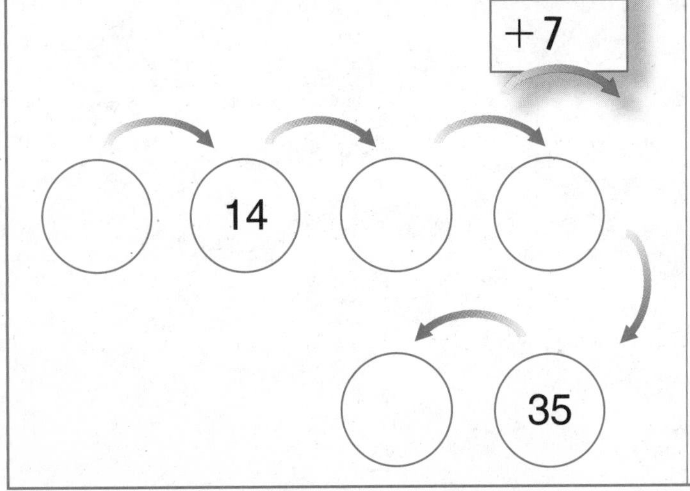

6. Write >, <, or =.

4 dimes _____ 50¢

3 quarters _____ 75¢

$1.00 _____ 11 dimes

Date Time

What Time Is It?

1. Write the time.

_____:_____ _____:_____ _____:_____ _____:_____

2. Draw the hands to match the time.

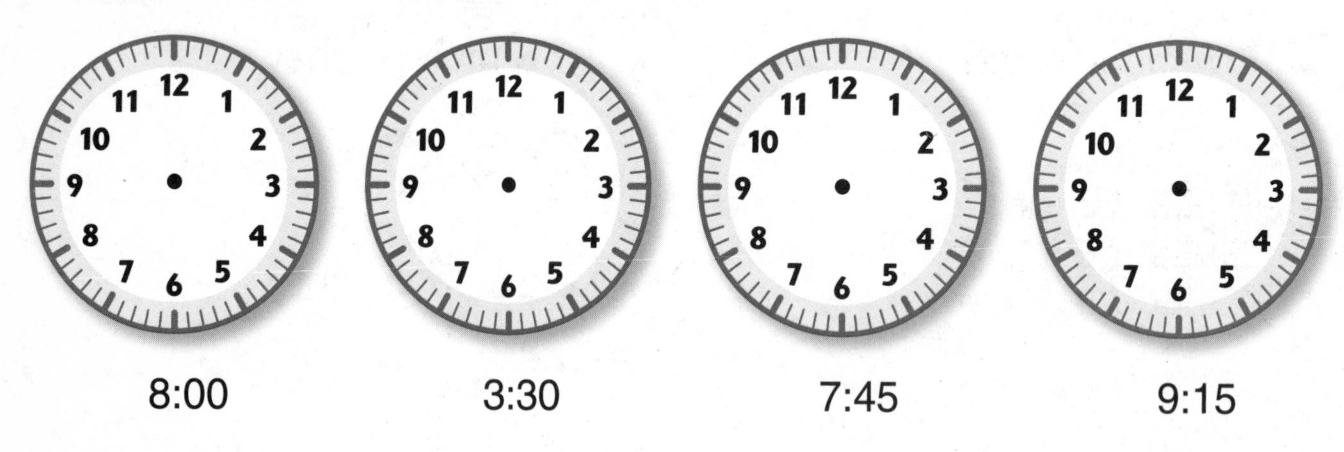

8:00 3:30 7:45 9:15

3. Make up times of your own. Draw the hands to show each time. Write the time under each clock.

_____:_____ _____:_____ _____:_____

Date Time

Frames and Arrows

1. Fill in the empty frames.

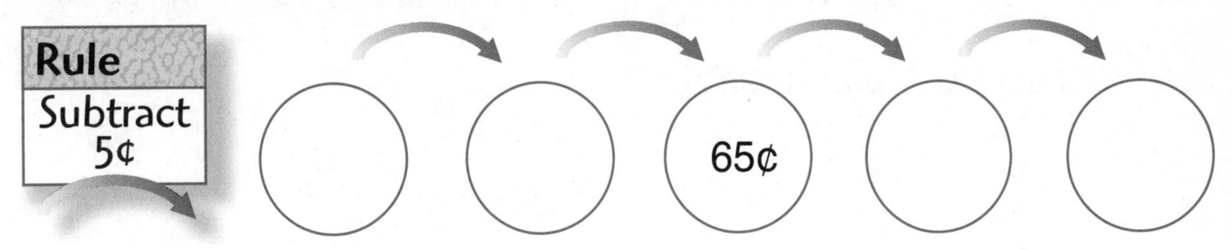

2. Fill in the arrow rule.

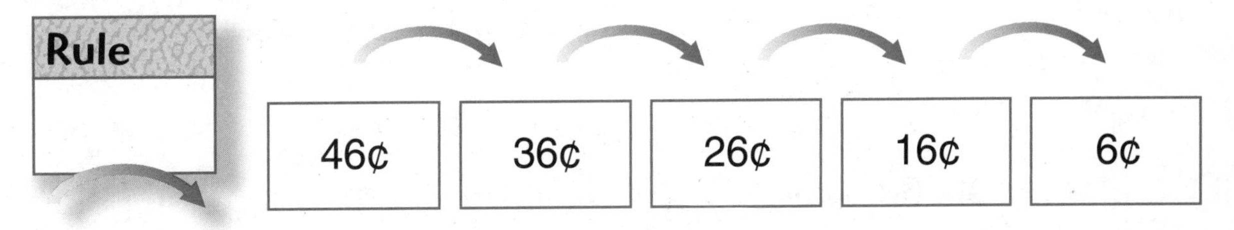

3. Fill in the arrow rule and the empty frames.

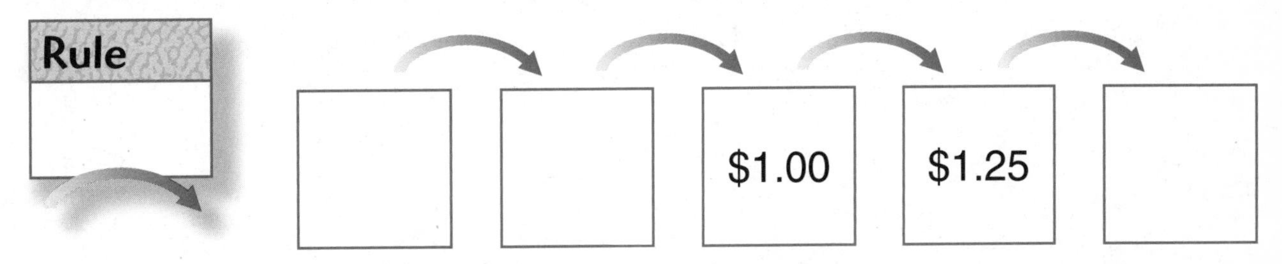

4. Make up your own Frames-and-Arrows problem.
Ask your partner to solve it.

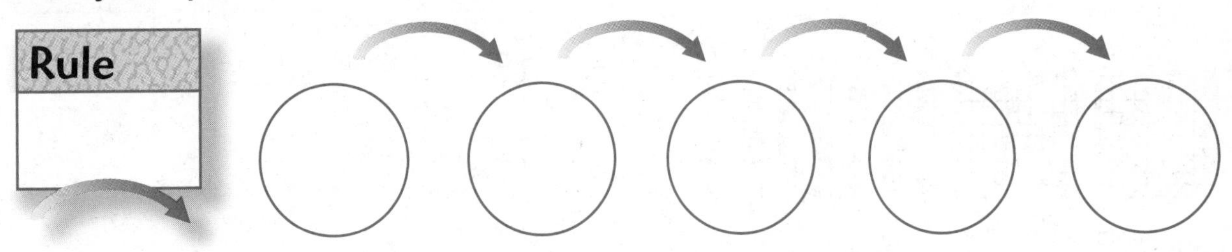

Date Time

Math Boxes 3.3

1. Use the digits 5 and 7 to make:

   the smallest possible number.

   ______

   the largest possible number.

   ______

2. 4 hundreds

   5 tens

   2 ones

   Write the number. ______

   Read it to yourself.

3. One plum costs 6¢. Buy 4 plums. How much money do you need?

   ______

4. Fill in the missing numbers.

Rule
+8

in	out
9	
6	
	18
13	

5. How many in all?

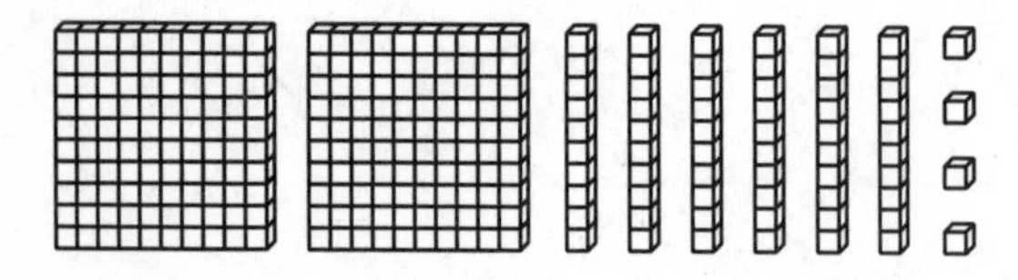

   ______

6. Solve.

Unit

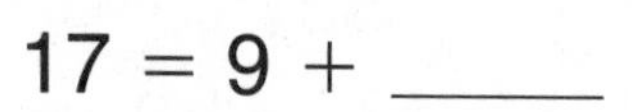

$18 = 9 +$ ______

$17 = 9 +$ ______

$16 = 9 +$ ______

$15 = 9 +$ ______

Date Time

Build a Number

Your number	Show your number using base-10 blocks	Show your number another way using base-10 blocks

Date Time

Geoboard Dot Paper (7 × 7)

1.

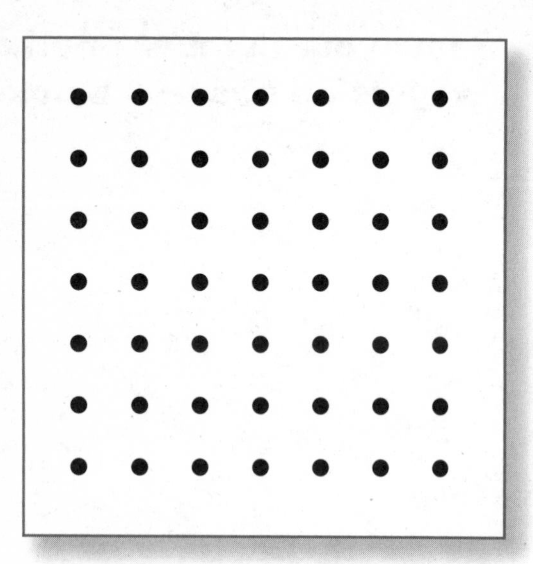

2.

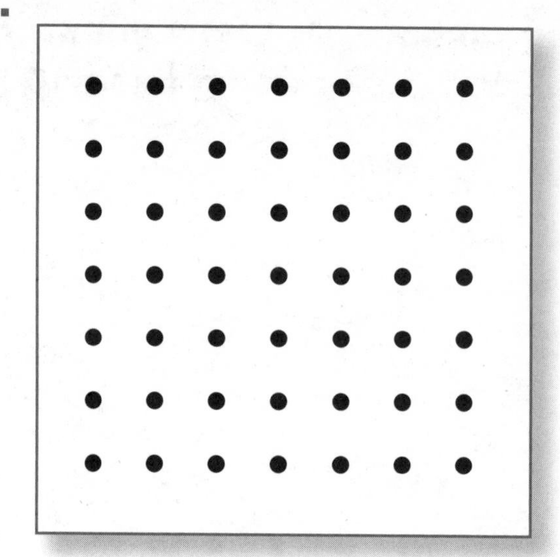

3.

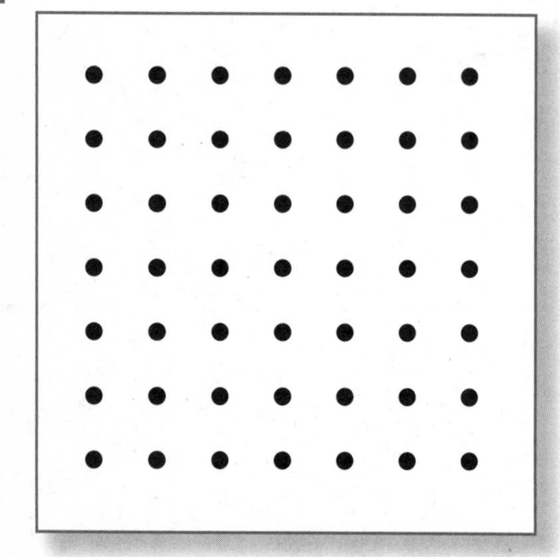

4.

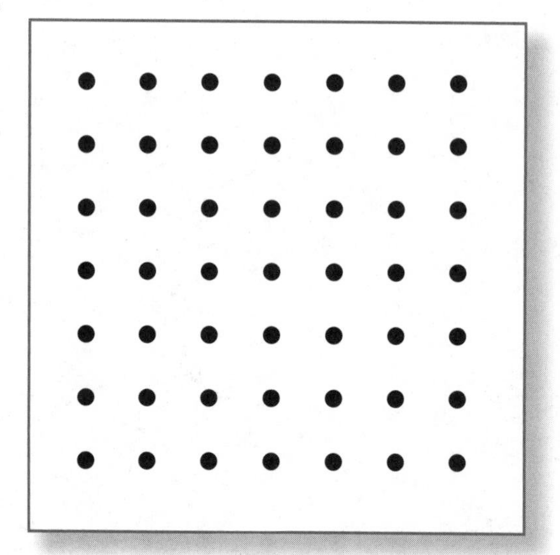

Dollar Rummy

Materials ❑ *Dollar Rummy* cards (*Math Masters,* p. 48)

❑ scissors to cut out cards

❑ cards from *Math Masters,* p. 49 for a harder game

Players 2

Directions

1. Deal 2 *Dollar Rummy* cards to each player.
2. Put the rest of the deck facedown between the players.
3. Take turns. When it's your turn, take the top card from the deck. Lay it faceup on the table.

4. Look for two cards that add up to $1.00. Use any cards that are in your hand or faceup on the table.
5. If you find two cards that add up to $1.00, lay these two cards facedown in front of you.

6. When you can't find any more cards that add up to $1.00, it is the other player's turn.
7. The game ends when all of the cards have been used or when neither player can make a $1.00 pair.
8. The winner is the player with more $1.00 pairs.

Date Time

Math Boxes 3.4

1. Draw the hands to show the time school begins.

2. Fill in the missing numbers.

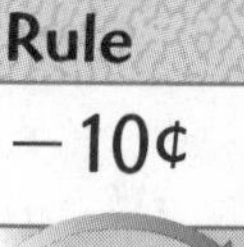

86¢

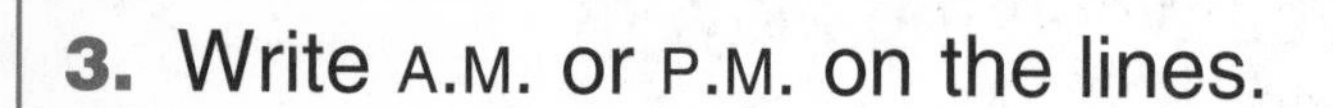

3. Write A.M. or P.M. on the lines.

7:00 ______ Wake up.

7:30 ______ Eat breakfast.

3:45 ______ Start homework.

6:00 ______ Eat dinner.

4. An apple costs 12¢ and a banana costs 9¢. Show the coins needed to buy both.

____

5. Use the digits 8 and 9 to make:

the smallest number possible.

______

the largest number possible.

______

6. Fill in the sum on the Fact Triangle. Write the fact family.

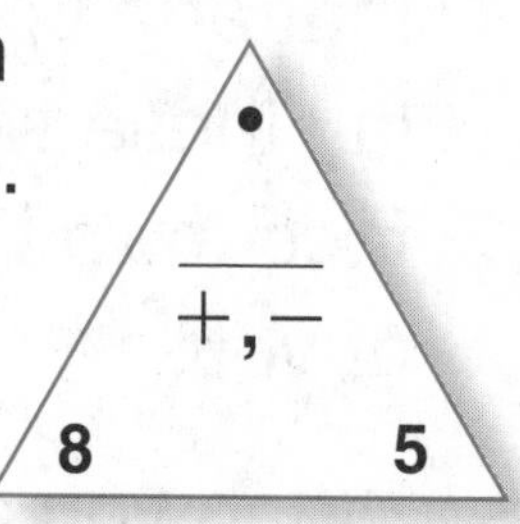

___ + ___ = ___

___ + ___ = ___

___ − ___ = ___

___ − ___ = ___

Date Time

Math Boxes 3.5

1. Solve.

Unit

100 = 50 + ____

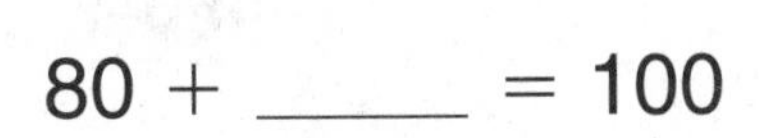

80 + ____ = 100

____ + 30 = 100

100 = 10 + ____

2. Write the number.

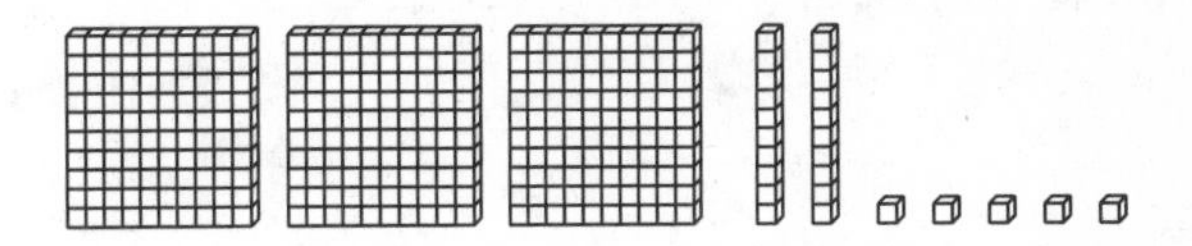

____

3. Fill in the missing numbers.

Rule
−3

in	out
	4
10	
	6
	5

4. Write 6 names in the 20-box.

20

5. Write the fact family.

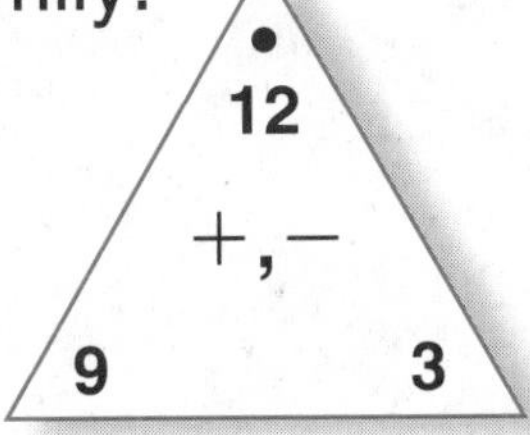

____ + ____ = ____

____ + ____ = ____

____ − ____ = ____

____ − ____ = ____

6. Solve.

Unit

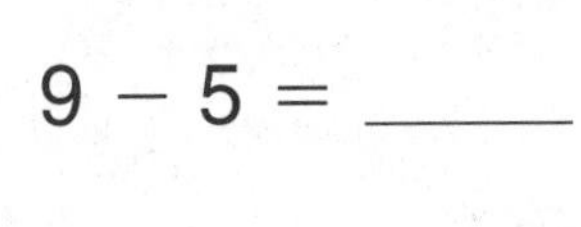

9 − 5 = ____

____ = 6 − 2

$$\begin{array}{r} 8 \\ -\ 1 \\ \hline \end{array} \qquad \begin{array}{r} 10 \\ -\ 6 \\ \hline \end{array}$$

Date Time

Pockets Data Table

Count the pockets of children in your class.

Pockets	Children	
	Tallies	Number
0		
1		
2		
3		
4		
5		
6		
7		
8		
9		
10		
11		
12		
13 or more		

Date Time

Graphing Pockets Data

Draw a bar graph of the pockets data.

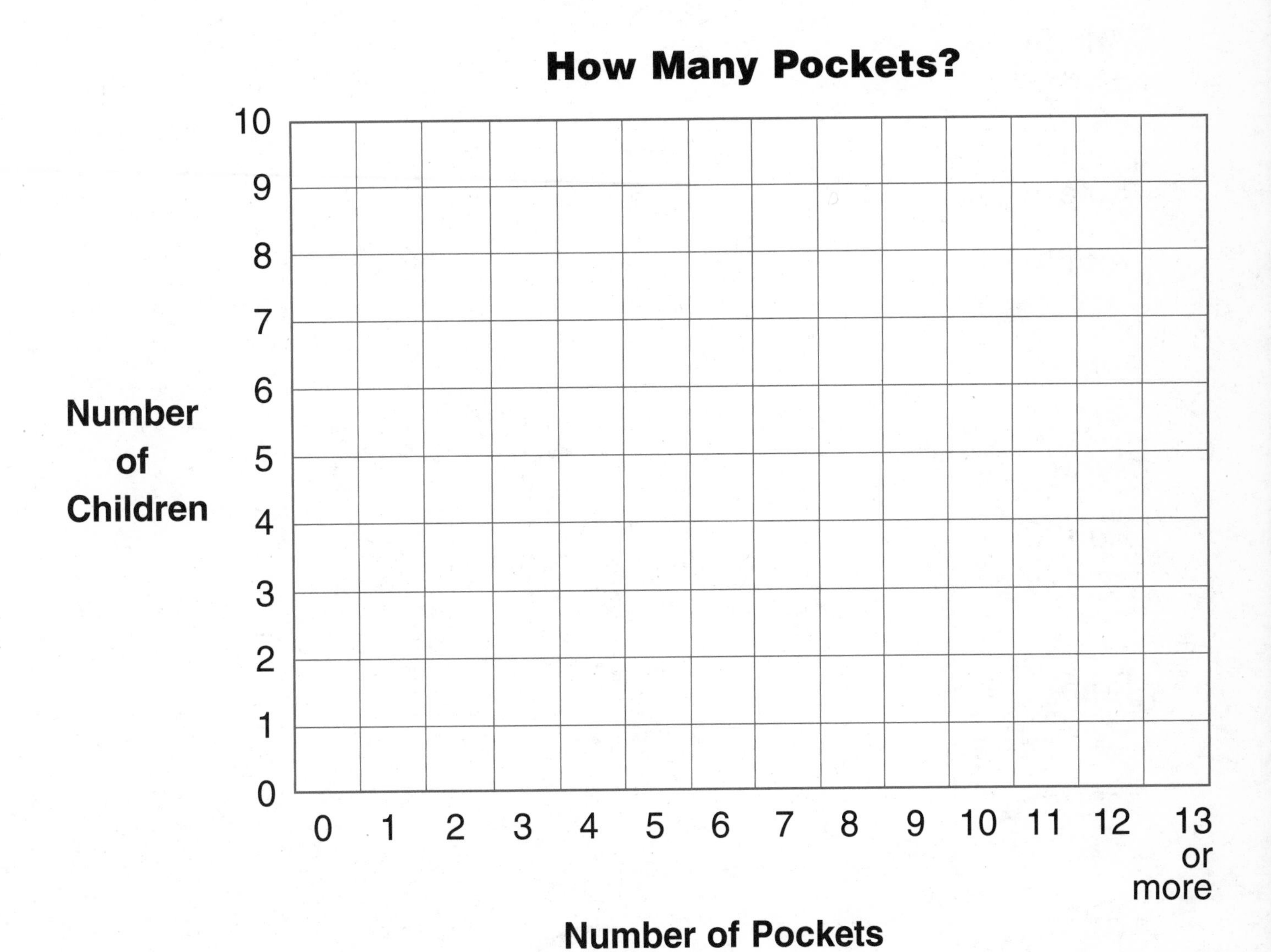

Date Time

Frames-and-Arrows Problems Having Two Rules

Fill in the frames. Use coins to help you.

1.

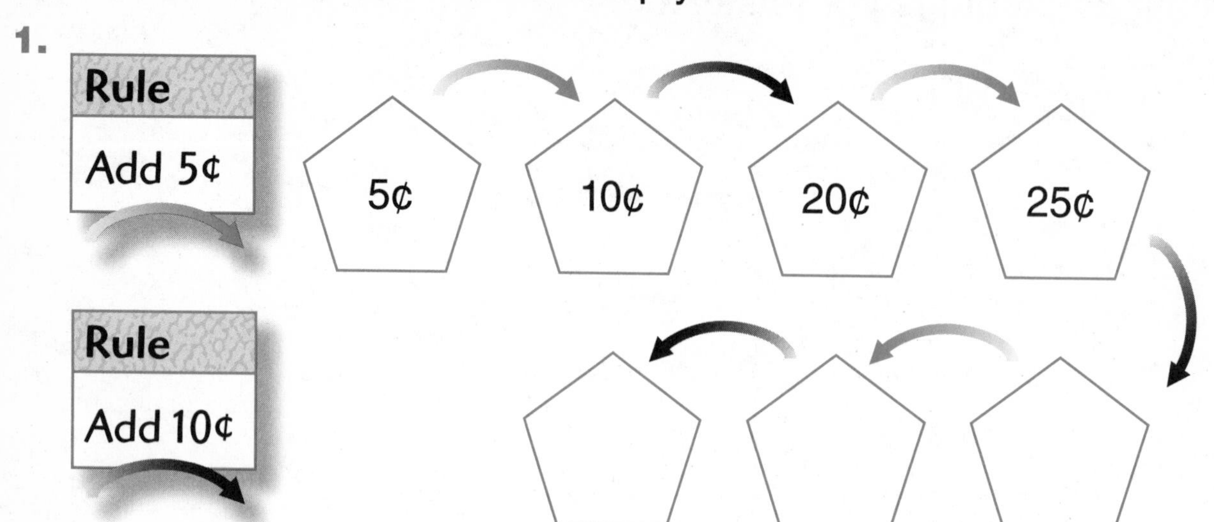

2.

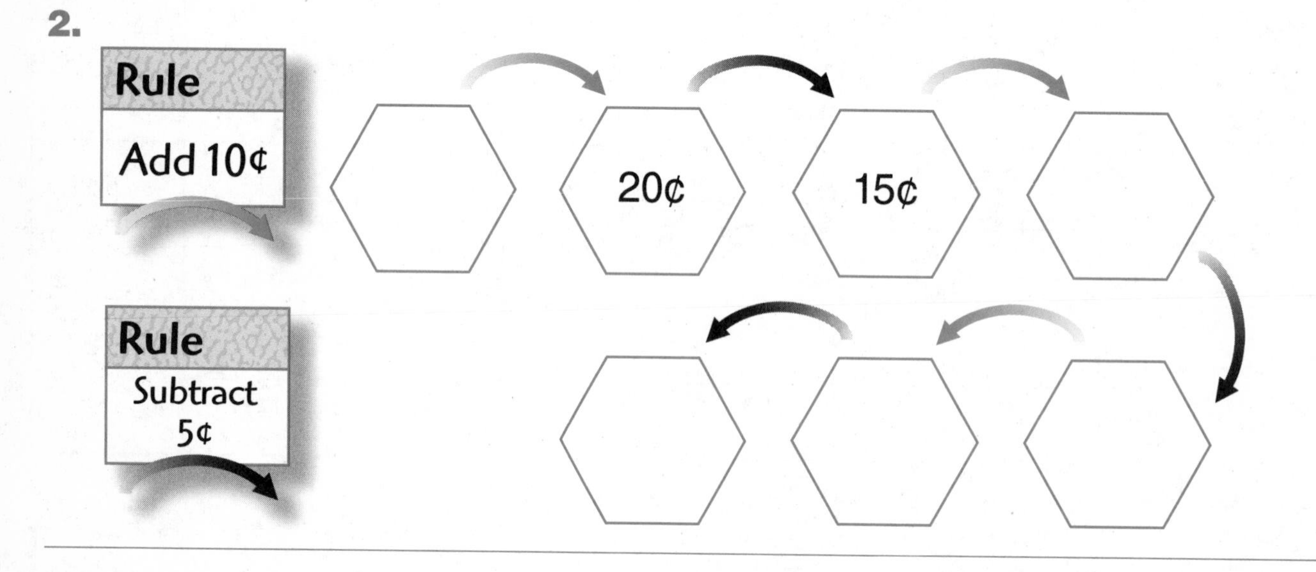

3.

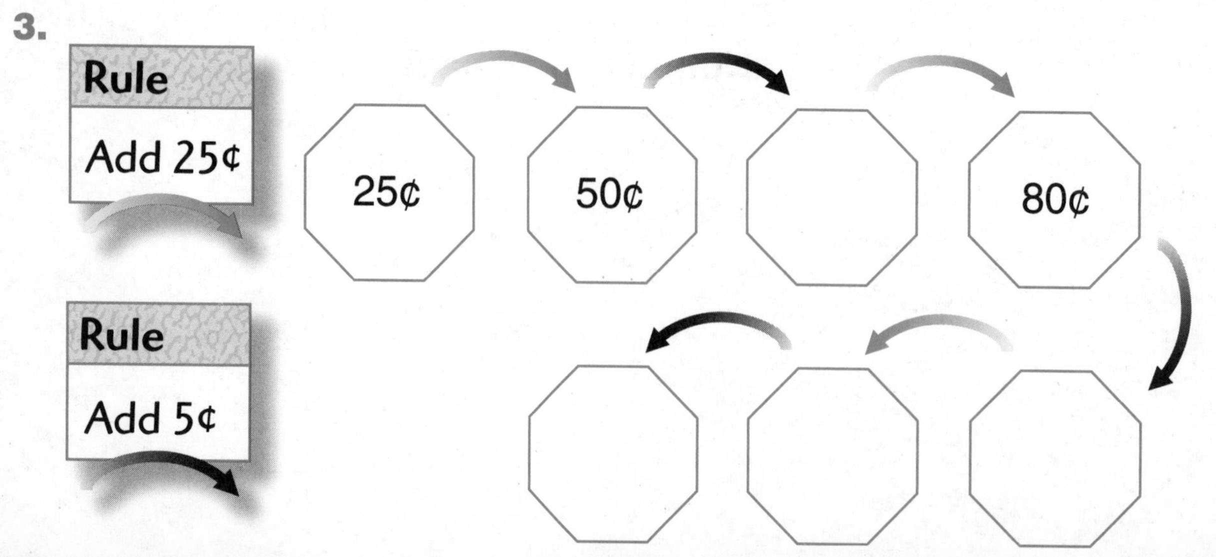

Use with Lesson 3.6.

Date Time

Frames-and-Arrows Problems (cont.)

4. Fill in the frames. Use coins to help you.

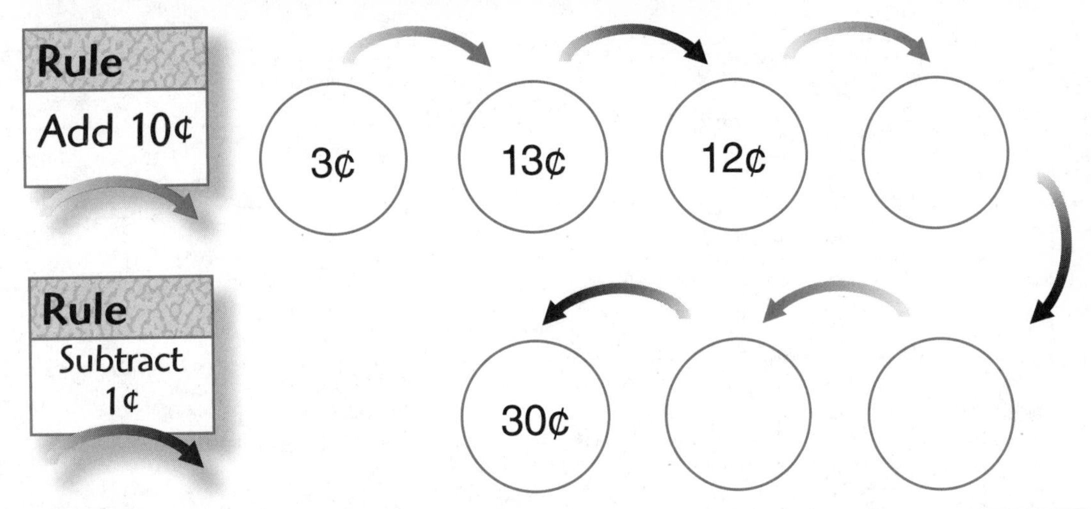

Fill in the frames and find the missing rules. Use coins to help you.

5.

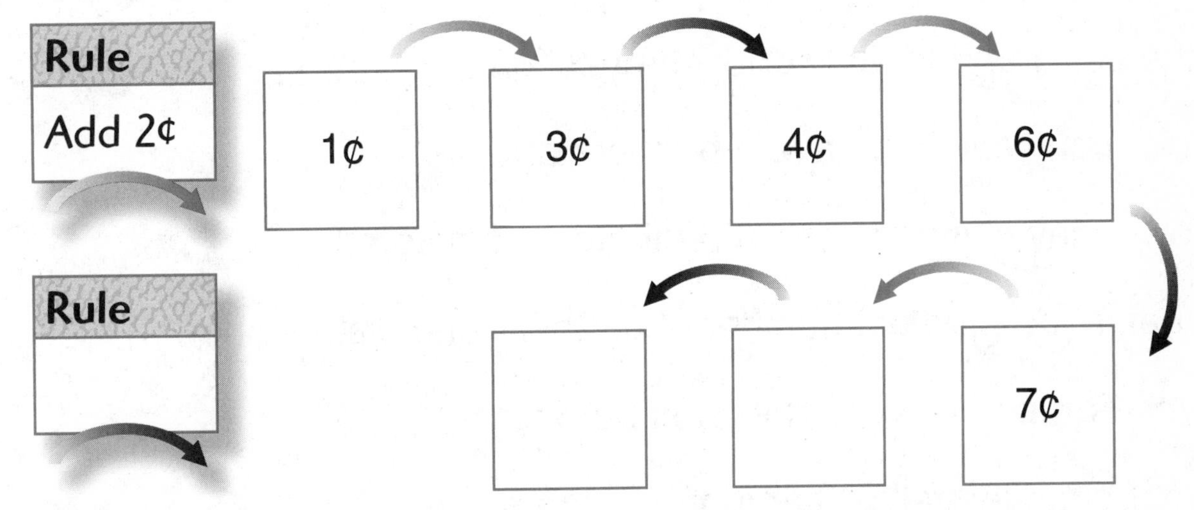

6.

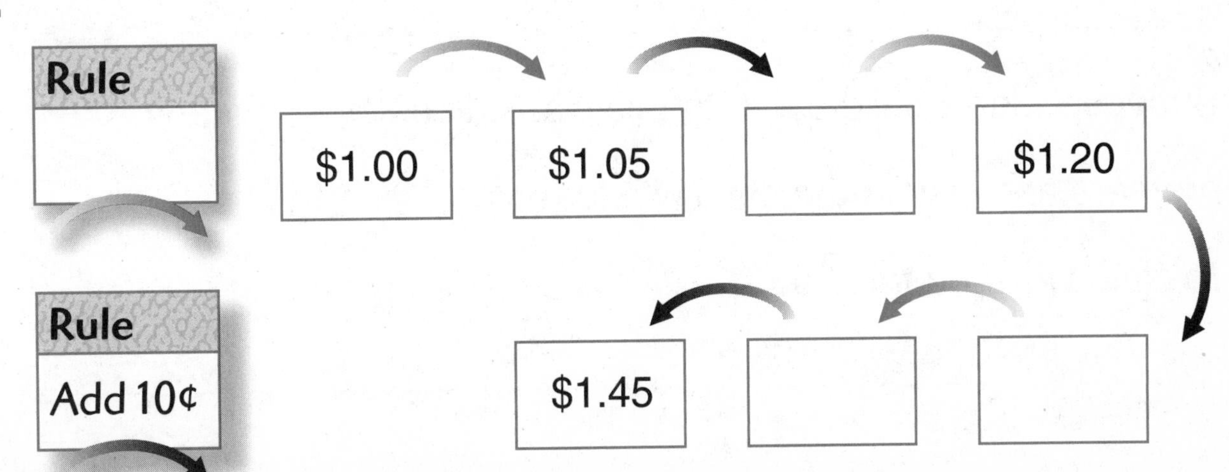

Date Time

Reading a Graph

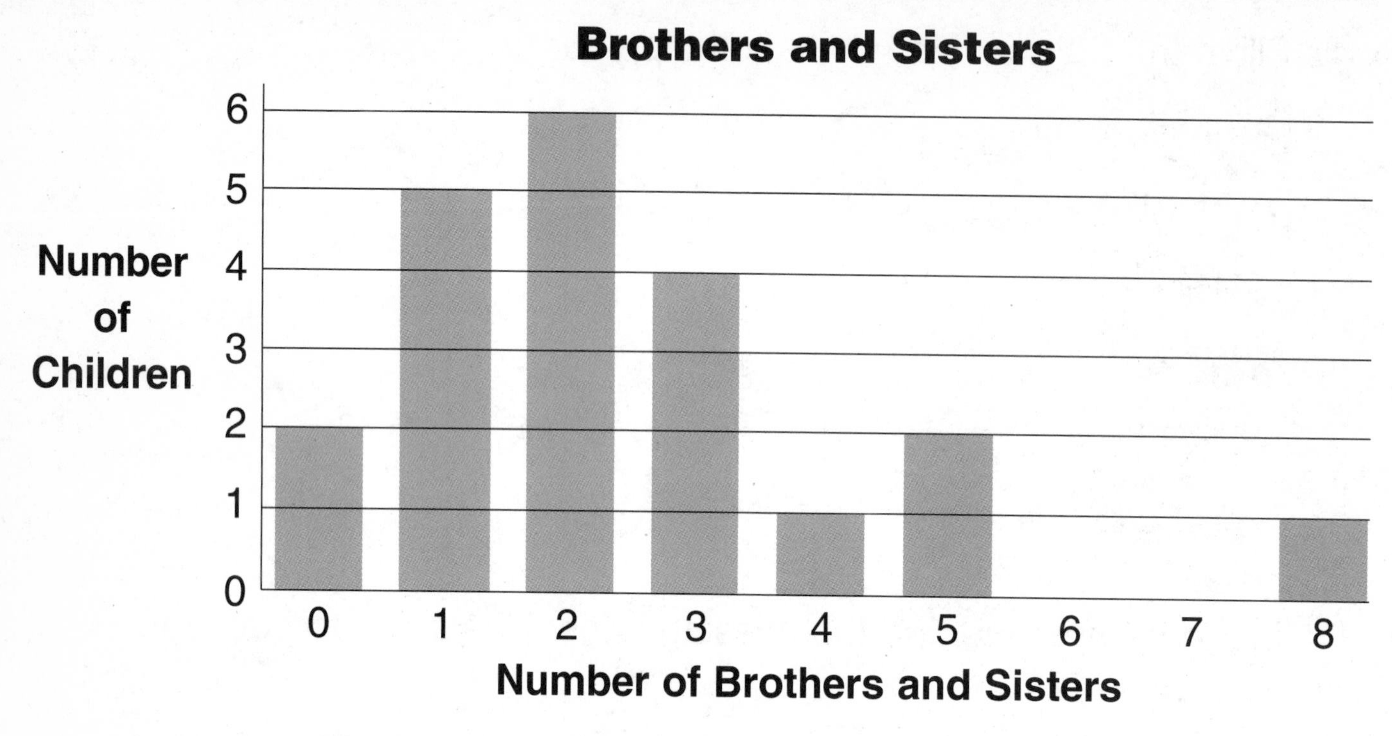

Use the bar graph to answer each question.

1. How many children have 5 brothers and sisters? ______

2. How many children have 3 brothers and sisters? ______

3. What is the greatest number of brothers and sisters shown by the graph? ______ The least number? ______

4. What does the tallest bar in the graph show?

____

5. Suppose a new child came to this class. Predict how many brothers and sisters the child would have. ______________

6. Explain why you think so.

____

____

Use with Lesson 3.6.

Date Time

Math Boxes 3.6

1. Write the time.

____ : ____

One hour later will be

____ : ____.

2. Fill in the frames.

Rule
+$0.25

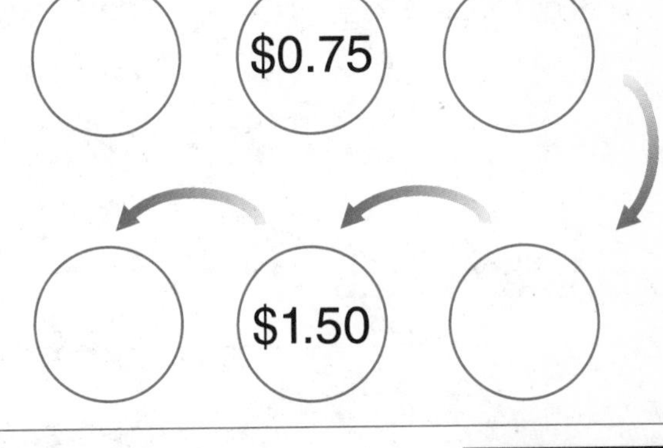

3. 843

There are ____ hundreds.

There are ____ tens.

There are ____ ones.

4. Solve.

Unit

____ = 10 − 3

10 − 5 = ____

$$\begin{array}{r} 10 \\ -\ 2 \\ \hline \end{array} \qquad \begin{array}{r} 10 \\ -\ 9 \\ \hline \end{array}$$

5. Marcus bought 9 stickers. Jean bought 4 stickers. How many more did Marcus buy?

____ stickers

Write a number model.

____

6. Show $0.88 two ways. Use Ⓠ, Ⓓ, Ⓝ, and Ⓟ.

____

____

____

Date Time

Shopping at the Fruit and Vegetables Stand

Price per Item					
pear	13¢	melon slice	30¢	lettuce	45¢
orange	18¢	apple	12¢	green pepper	24¢
banana	9¢	tomato	20¢	corn	15¢
plum	6¢	onion	7¢	cabbage	40¢

Complete the table.

I bought	I paid (Draw coins or $1 bill)	I got in change
______		______ ¢
______		______ ¢
______		______ ¢

Challenge

Buy 2 items. How much change from $1.00 will you get?

I bought	I paid	I got in change
______ and ______	$1	______ ¢

Date Time

Frames-and-Arrows Problems Having Two Rules

Fill in the frames. Use coins to help you.

1.

Rule

Add 10¢

30¢ → 40¢ → 47¢ → 57¢

Rule

Add 7¢

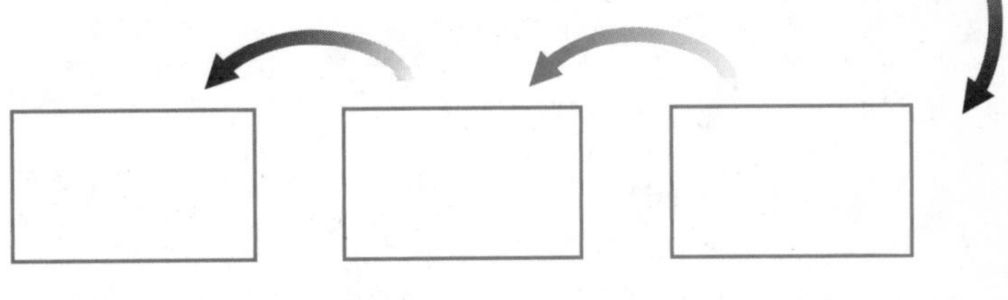

2.

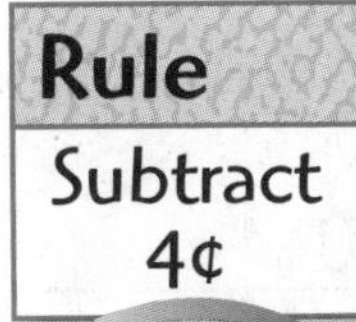

Rule

Subtract 4¢

20¢ → 16¢ → 26¢ → 22¢

Rule

Add 10¢

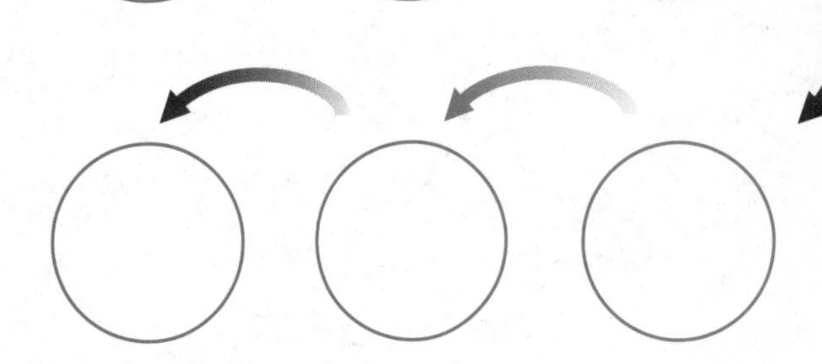

3.

Rule

Subtract 10¢

$1.20 → $1.10 → $1.02 → 92¢

Rule

Subtract 8¢

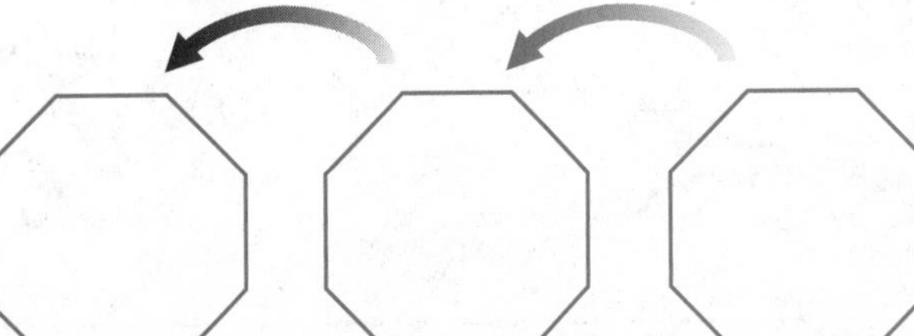

Use with Lesson 3.7.

Math Boxes 3.7

1. Solve.

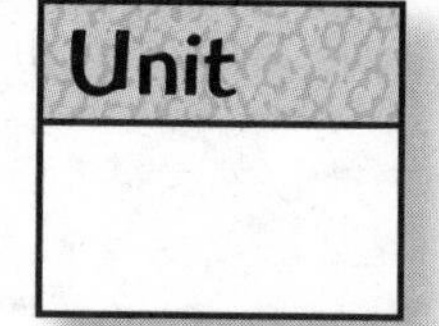

$15 > 8 + \_\_\_\_$

$6 + 7 < 6 + \_\_\_\_$

$17 = 9 + \_\_\_\_$

2. How much money?

$ ________

3. Draw hands on the clock to show 6:15.

4. Fill in the missing numbers.

Rule
Add 1 hour

in	out
6:00	
12:30	
	3:15
	1:45

5. Write the fact family.

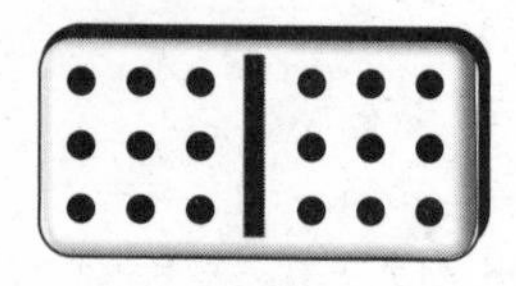

____ + ____ = ____

____ − ____ = ____

6.

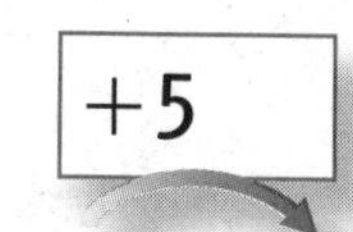

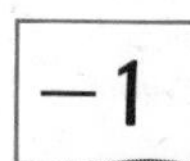

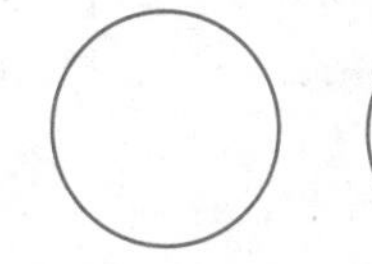

3 → ( ) → ( ) → 12

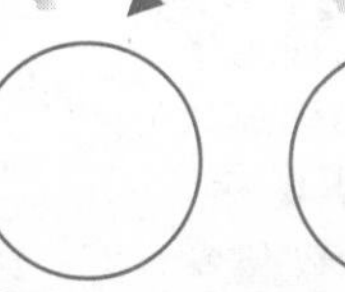
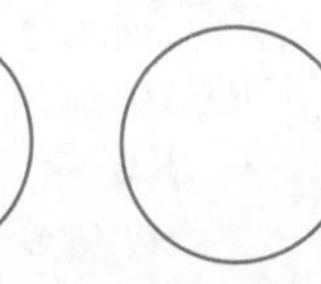

20 ← ( ) ← ( ) ← ( )

Date Time

Math Boxes 3.8

1.

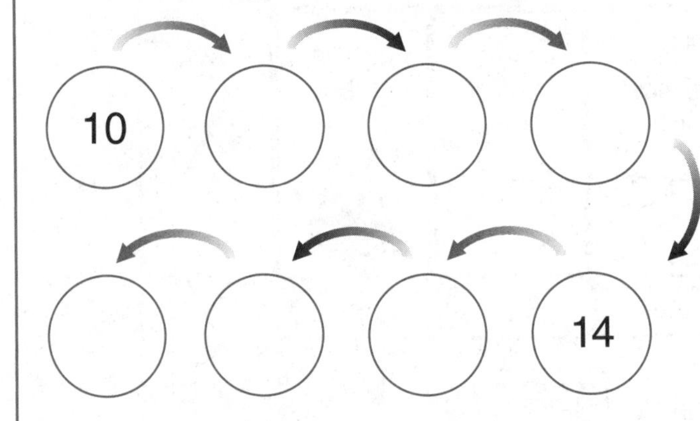

2. What time is it?

___ : ___

What time was it 2 hours earlier?

___ : ___

3. The total cost is 16¢. I pay with 2 dimes. How much change do I get?

___

4. Solve.

$$\begin{array}{r} 9 \\ -\ 3 \\ \hline \end{array} \qquad \begin{array}{r} 8 \\ -\ 5 \\ \hline \end{array}$$

Unit

___ = 7 + 6

___ = 5 + 8

5. Write 3 names for 50.

6. 4 ones

3 hundreds

7 tens

Write the number. (Be careful!)

___

Date Time

Buying from a Vending Machine

1. The exact change light is on. You want to buy a carton of orange juice. Which coins will you put in? Draw Ⓝ, Ⓓ, and Ⓠ to show the coins.

2. The exact change light is off. You want to buy a carton of 2% milk. You don't have the exact change. Which coins or bills will you put in? Draw coins or a $1 bill.

   How much change will you get? _______________

Date Time

Buying from a Vending Machine (cont.)

3. The exact change light is on.

You buy:	Draw the coins you put in.
chocolate milk	
yogurt drink	
______________	

4. The exact change light is off.

You buy:	Draw the coins or the $1 bill you put in.	What is your change?
orange juice	Ⓠ Ⓠ Ⓠ	_____¢
chocolate milk	$1	_____¢
______________		_____¢
______________		_____¢
______________		_____¢

Date Time

Making Change

Materials
- ❑ 2 nickels, 2 dimes, 2 quarters, and one $1 bill for each player
- ❑ 2 six-sided dice
- ❑ a cup, a small box, or a piece of paper to use as a bank

Players 2 or 3

Directions

1. Each player starts the game with 2 nickels, 2 dimes, 2 quarters, and one $1 bill. Players take turns rolling the dice and finding the total number of dots that are faceup.
2. Players use the chart to find out how much money they must put in the bank. (There is no money in the bank at the beginning of the game.)

Making Change Chart

Total for Dice Roll	2	3	4	5	6	7	8	9	10	11	12
Amount to Pay the Bank	10¢	15¢	20¢	25¢	30¢	35¢	40¢	45¢	50¢	55¢	60¢

3. Players use their coins to pay the amount to the bank. Players can get change from the bank.
4. The winner is the first player who doesn't have enough money left to pay the bank.

Date Time

Math Boxes 3.9

1. Solve.

Unit

100 = 75 + _____

60 + _____ = 100

$$\begin{array}{r} 95 \\ +\ \ 5 \\ \hline \end{array} \qquad \begin{array}{r} 90 \\ +\ 10 \\ \hline \end{array}$$

2. 45 cents = 1 quarter

and _____ dimes

60 cents = 3 dimes

and _____ nickels

3. Use the digits 1, 3, and 5 to make:

the smallest number possible.

_____

the largest number possible.

_____

4. Lea found 1 nickel and 3 pennies. Jake found 1 dime and 1 nickel. How much money did they find in all?

_____

5. Complete the Fact Triangle. Write the fact family.

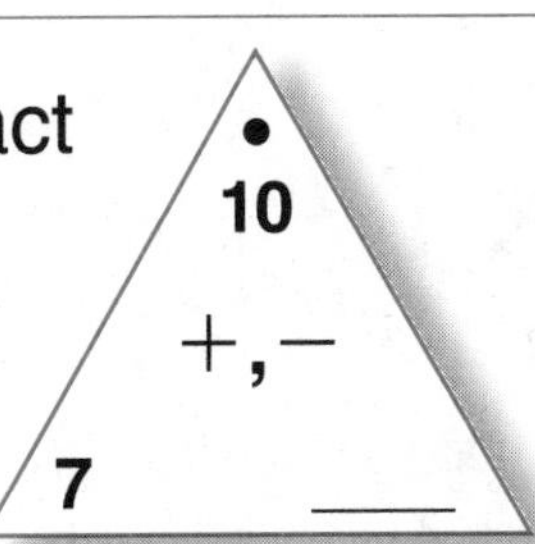

___ + ___ = ___

___ + ___ = ___

___ − ___ = ___

___ − ___ = ___

6. The cost of a piece of candy is 11¢. I pay with 15¢. How much change do I get?

_____

Date Time

Fish Poster

Fish A
1 lb
12 in.

Fish B
3 lb
14 in.

Fish C
4 lb
18 in.

Fish D
5 lb
24 in.

Fish E
6 lb
24 in.

Fish F
8 lb
30 in.

Fish G
10 lb
30 in.

Fish H
14 lb
30 in.

Fish I
15 lb
30 in.

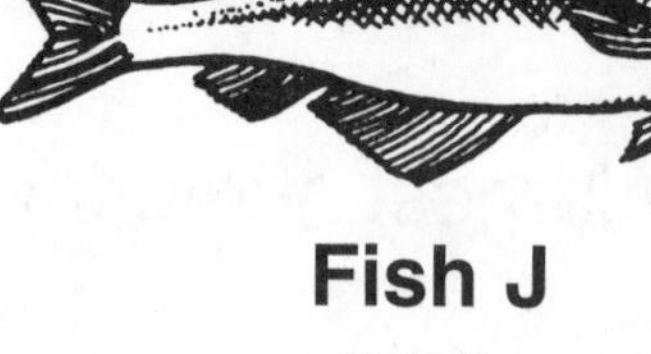

Fish J
24 lb
36 in.

Fish K
35 lb
42 in.

Fish L
100 lb
72 in.

Date Time

"Fishy" Stories

Use the information on journal page 80 for Problems 1–4.

Do the following for each number story:

- Write the numbers you know in the change diagram.
- Write ? for the number you need to find.
- Answer the question.
- Write a number model.

---

1. Fish J swallows Fish B.

   How much does Fish J weigh now?

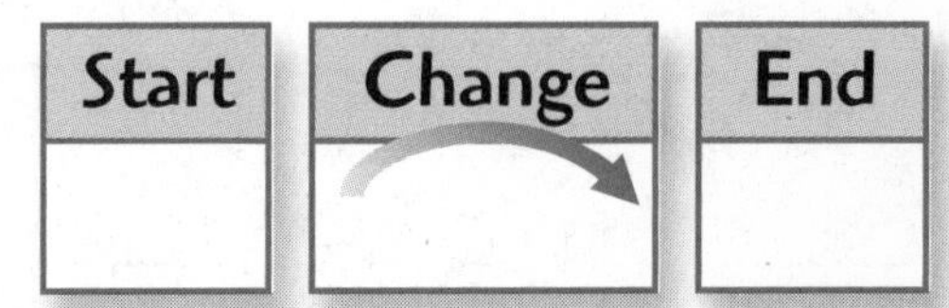

   Answer: ______ pounds

   Number model: ____________________

---

2. Fish K swallows Fish C.

   How much does Fish K weigh now?

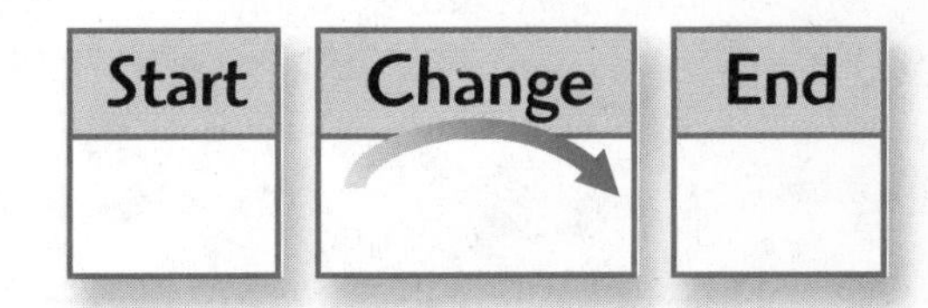

   Answer: ______ pounds

   Number model: ____________________

---

3. Fish L swallows Fish F.

   How much does Fish L weigh now?

   Answer: ______ pounds

   Number model: ____________________

Date Time

"Fishy" Stories (cont.)

4. Fish I swallows another fish.

Fish I now weighs 18 pounds.

Which fish did Fish I swallow?

Start	Change	End

Answer: ________________

Number model: ____________________

Write the missing number in each change diagram.

5.

Start	Change	End
30	+50	

6.

Start	Change	End
20	+6	

7.

Start	Change	End
40		45

8.

Start	Change	End
9		89

9.

Start	Change	End
	+30	39

10.

Start	Change	End
	+3	13

 Use with Lesson 4.1.

Date Time

Distances on a Number Line

Example

Jump by 1s from 47 to 56. How many jumps? 9

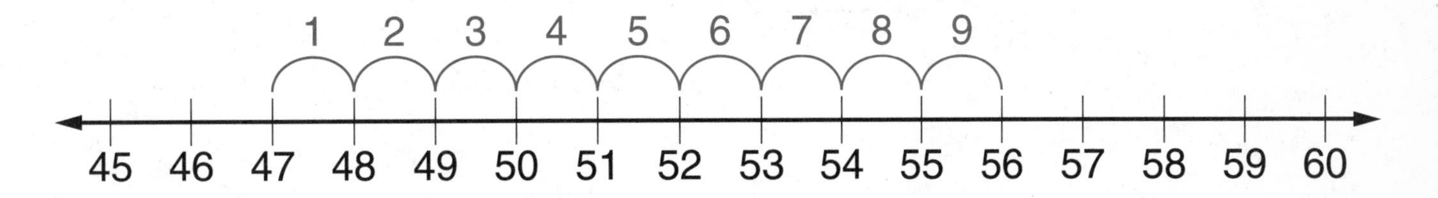

1. Jump by 1s from 34 to 22. How many jumps? ______

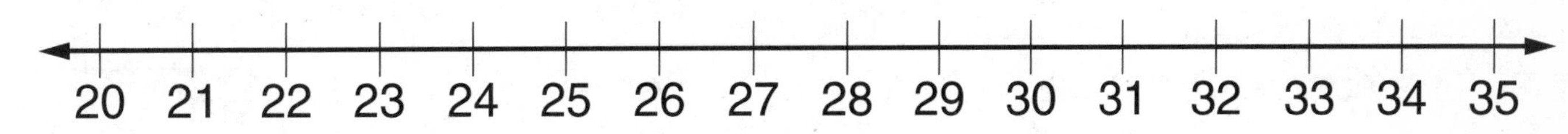

2. How many jumps by 1s from 69 to 84? ______

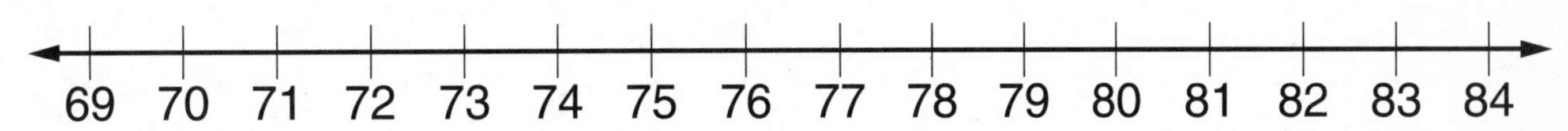

3. Jump by 1s from 39 to 28. How many jumps is that? ______

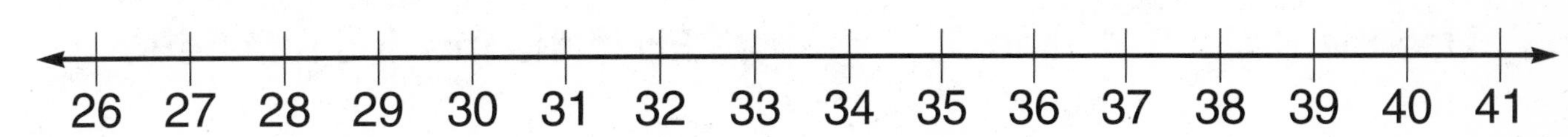

4. Jump by 1s from 65 to 77. How many jumps is that? ______

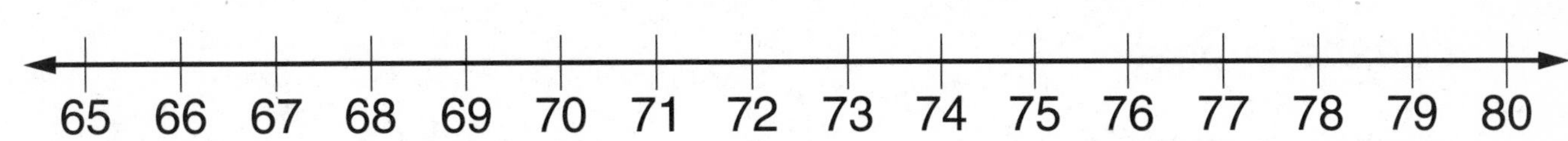

Date Time

Math Boxes 4.1

1. Solve.

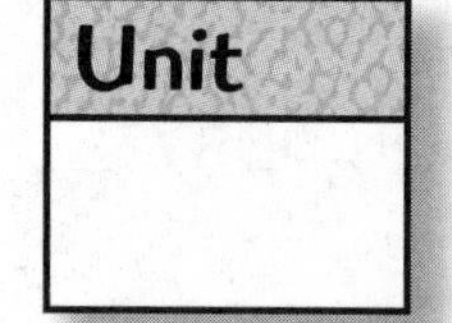

70 + _____ = 100

100 = 20 + 30 + _____

_____ = 50 + 40 + 10

100 = _____ + 10 + 60

2. Write the number. Be careful!

5 ones

6 hundreds

3 tens

_____

3. The cost is 18¢. I pay with a quarter. How much change do I get?

_____

4. Solve.

Unit

18 = 8 + _____

17 = 8 + _____

16 = 8 + _____

15 = 8 + _____

5. Use the digits 3, 1, and 5.

Write the smallest possible number. _____

Write the largest possible number. _____

6. Fill in the missing numbers.

	89	
		100

Date Time

Parts-and-Total Number Stories

Lucy's Snack Bar Menu					
Sandwiches		**Drinks**		**Desserts**	
Hamburger	65¢	Juice	45¢	Apple	15¢
Hot dog	45¢	Milk	35¢	Orange	25¢
Cheese	40¢	Soft drink	40¢	Banana	10¢
Peanut butter and jelly	35¢	Water	25¢	Cherry pie	40¢

For Problems 1–4, you are buying two items. Use the diagrams to record both the cost of each item and the total cost.

1. a cheese sandwich and milk

Total	
Part	Part

2. juice and a slice of pie

Total	
Part	Part

3. a hot dog and an apple

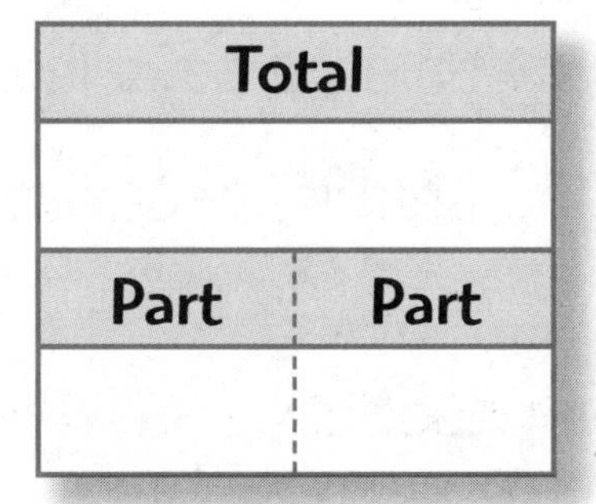

4. a hamburger and juice

Total	
Part	Part

5. Jean buys milk and an orange. The cost is ______.

Jean gives the cashier 3 quarters.

How much change does she get? ______

Addition Spin

Materials ❑ paper clip ❑ pencil

❑ *Math Masters,* p. 62

❑ calculator

❑ 2 sheets of paper

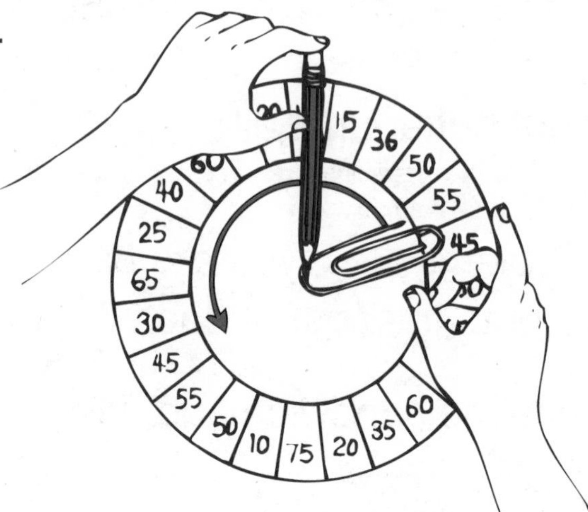

1. Players take turns being the "Spinner" and the "Checker."
2. The Spinner uses a pencil and a paper clip to make a spinner on *Math Masters,* page 62.
3. The Spinner spins the paper clip.
4. The Spinner writes the number that the paper clip points to. If the spinner points to more than one number, the Spinner writes the smaller number.
5. The Spinner spins a second time and writes the new number.
6. The Spinner adds the two numbers and writes the sum. The Checker checks the sum of the two numbers by using a calculator.
7. If the sum is correct, the Spinner circles it. If the sum is incorrect, the Spinner corrects it but does not circle it.
8. Players switch roles. The new Spinner spins the paper clip and writes the numbers and their sum on another sheet of paper. The new Checker checks the sum.
9. Players stop after they have each played 5 turns. Each player uses a calculator to find the sum of his or her circled scores.
10. The player with the higher total wins.

Date Time

Math Boxes 4.2

1. Solve.

Unit

13 − 5 = _____

13 = 8 + _____

7 = 16 − _____

7 + _____ = 16

2. Write the total number of coins needed to make 67¢.

67¢ = _____ quarters

_____ dimes

_____ nickels

_____ pennies

3. Use a number line. Jump by 1s from 25 to 47. How many jumps?

_____ jumps

4. Mike had 7¢. He found a dime. How much money did he have then? ____¢ Fill in the diagram and write a number model.

Start	Change	End
	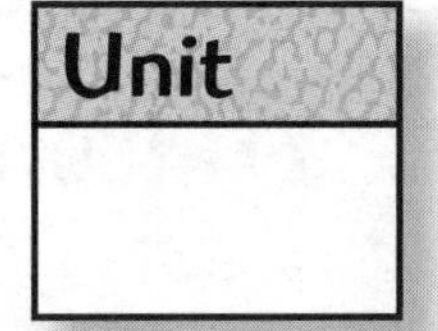	

____

5. Solve.

Unit

_____ = 70 − 60

_____ = 88 − 8

63 − 20 = _____

_____ = 92 − 30

6. Fill in the rule and the missing numbers.

Rule

in	out
132	122
103	93
114	
205	

Date Time

Temperatures

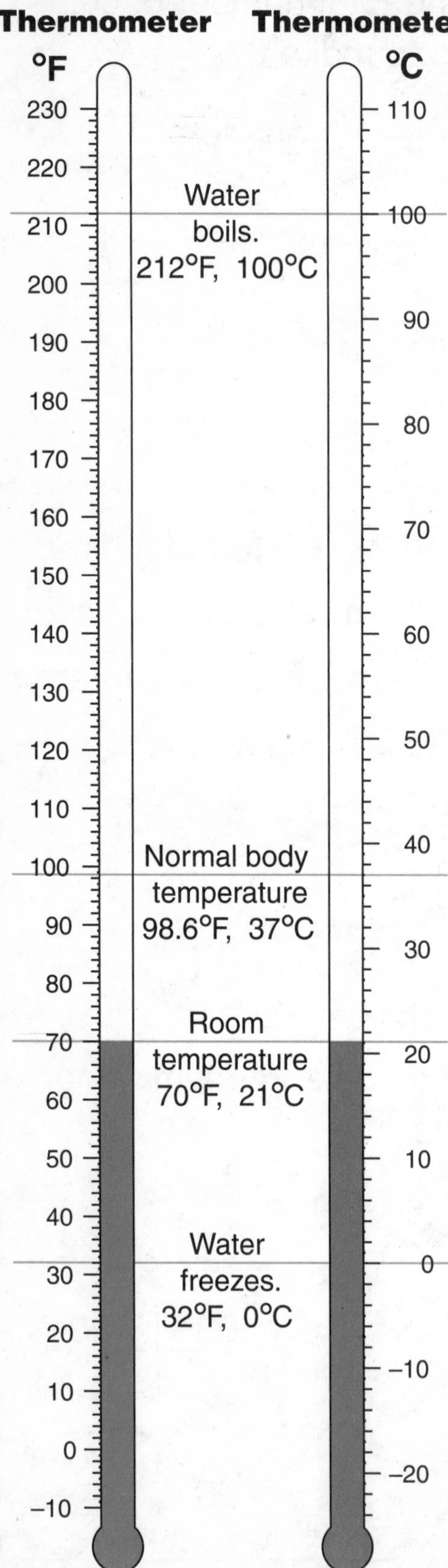

1. Use a thermometer to measure and record the temperatures of the following:

a. your classroom ______°F ______°C

b. hot water from a faucet ______°F ______°C

c. ice water ______°F ______°C

---

2. Which temperature makes more sense? Circle it.

a. temperature in a classroom:

40°F or 70°F

b. temperature of hot tea:

100°F or 180°F

c. temperature of a person with a fever:

100°F or 100°C

d. temperature on a good day for ice skating outside:

−10°C or 10°C

 Use with Lesson 4.3.

Date Time

Math Boxes 4.3

1. The total cost is 75¢. I pay with $1.00. How much change do I get?

_____

2. Fill in the frames.

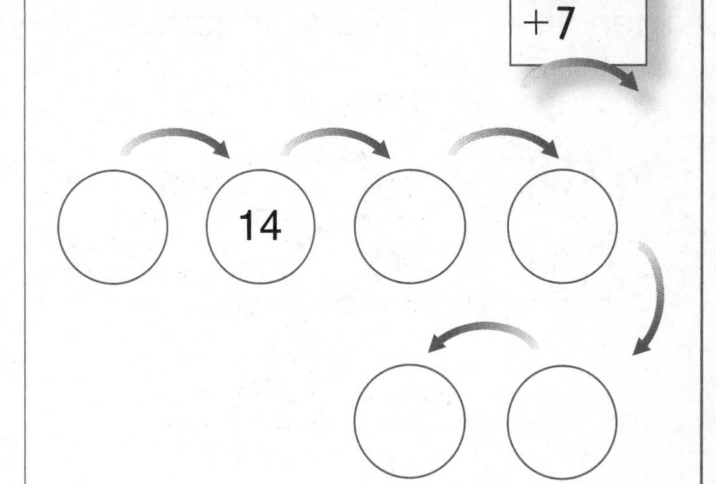

3. Circle the digits in the hundreds place.

1 2 8

9 7 2

4 6 4

2, 4 2 5

4. Had $30. Earned $17 more.

How much money now? $_____

Fill in the diagram and write a number model.

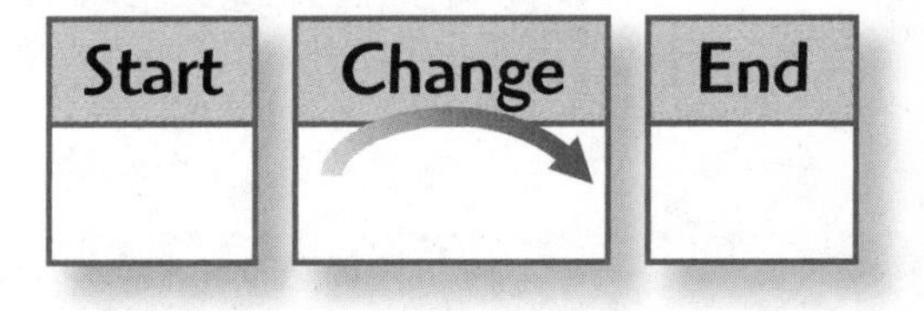

_____

5. How much?

Ⓠ Ⓠ Ⓠ Ⓓ Ⓓ Ⓝ Ⓟ

_____

6. Use a number line. Count by 1s. How many jumps from 84 to 60?

_____ jumps

Date Time

Temperature Changes

Write the missing number in each End box.

Then fill in the End thermometer to show this number.

Unit
°F

1.

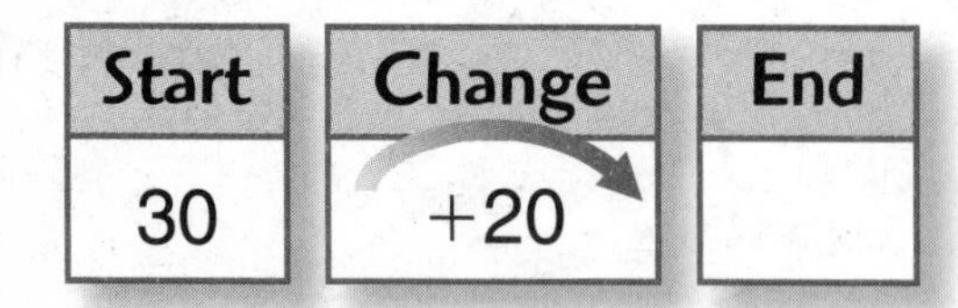

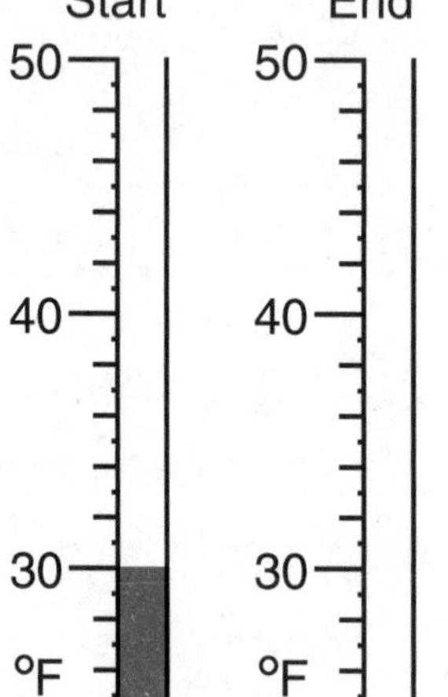

2.

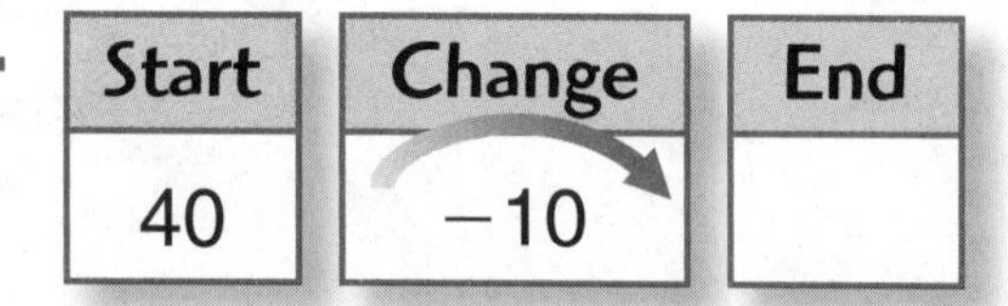

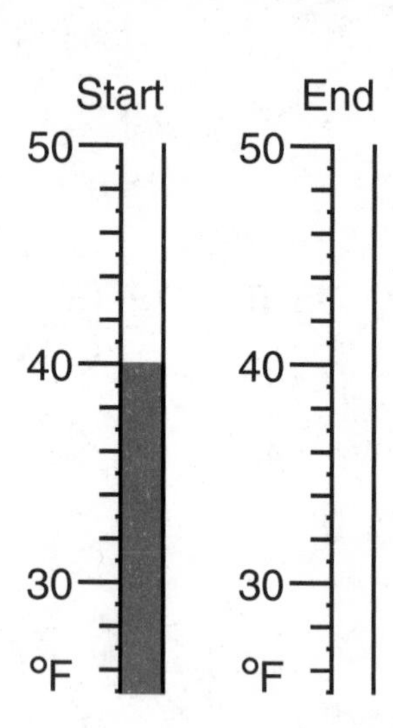

3.

Start	Change	End
44	−10	

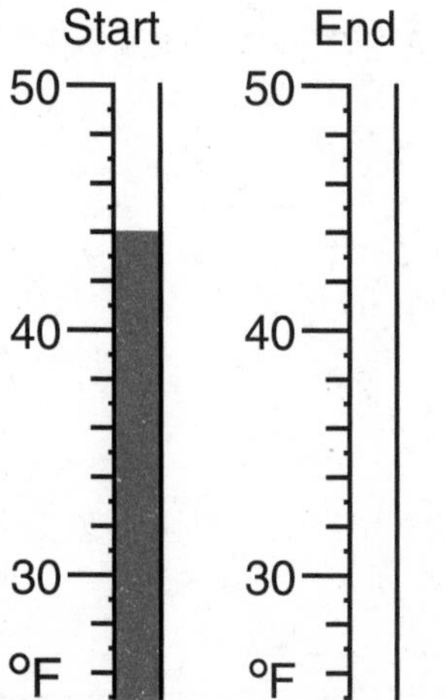

4.

Start	Change	End
70	−20	

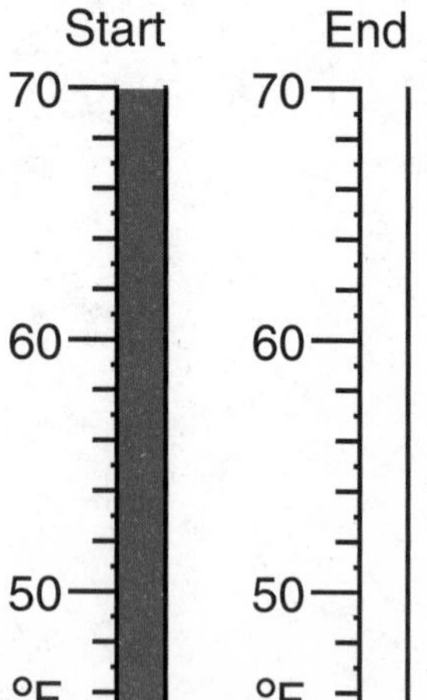

Date Time

Temperature Changes (cont.)

Fill in the missing numbers for each diagram.

5.

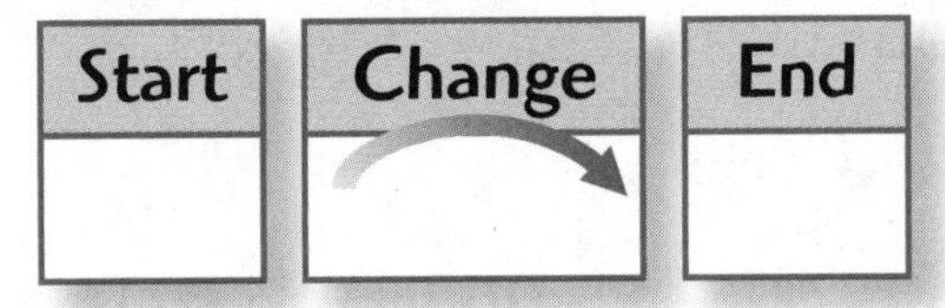

Start End

60 60

50 50

40 40

°F °F

6.

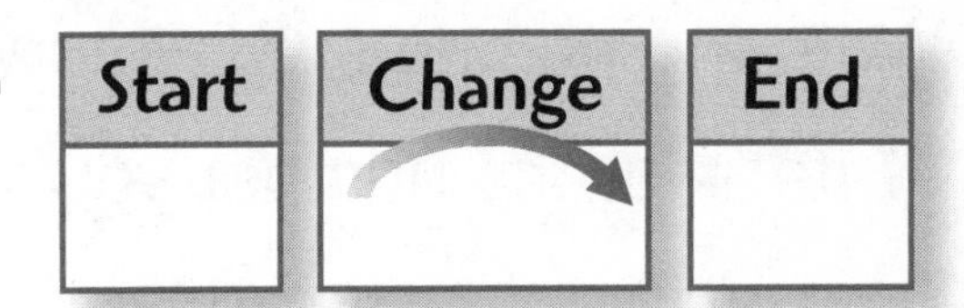

Start End

80 80

70 70

60 60

°F °F

7.

Start End

60 60

50 50

40 40

°F °F

8.

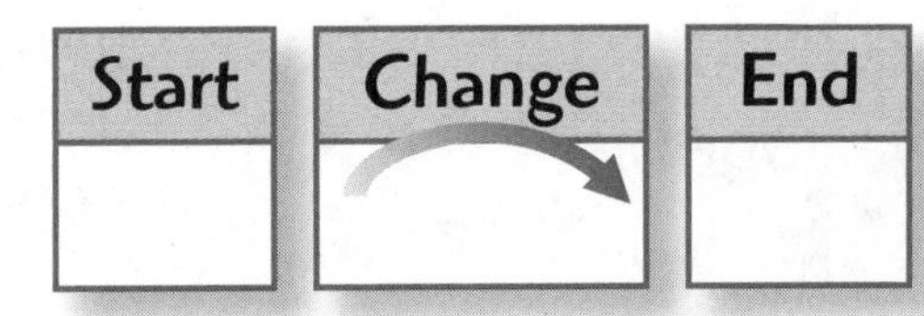

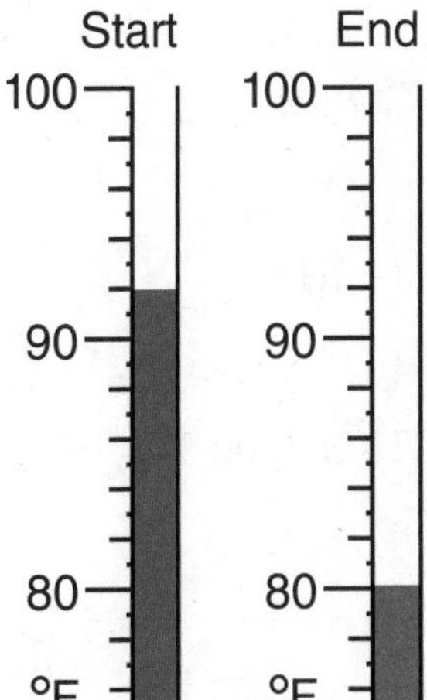

Parts-and-Total Number Stories

For each number story:

- Write the numbers you know in the parts-and-total diagram.
- Write ? for the number you want to find.
- Answer the question. Remember to include the unit.
- Write a number model.

1. Jack rode his bike for 20 minutes on Monday. He rode it for 30 minutes on Tuesday. How many minutes did he ride his bike in all?

Total	
Part	Part

Answer: ______________________
(unit)

Number model: ______________________

2. Two children collect stamps. One child has 40 stamps. The other child has 9 stamps. How many stamps do the two children have together?

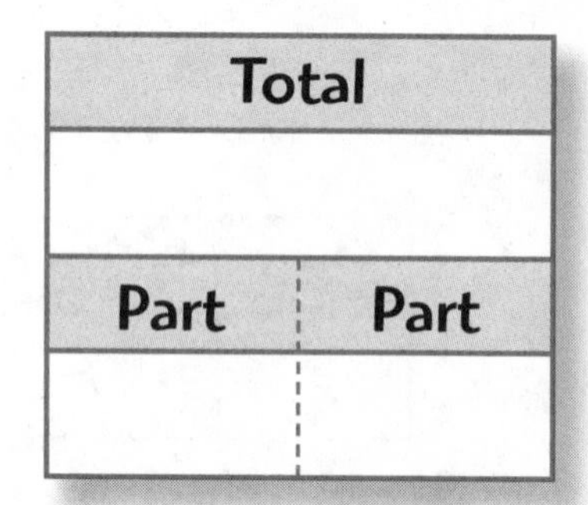

Answer: ______________________
(unit)

Number model: ______________________

3. 25 children take ballet class. 15 children take painting class. In all, how many children take the two classes?

Total	
Part	Part

Answer: ______________________
(unit)

Number model: ______________________

Date Time

Math Boxes 4.4

1. Fill in the frames.

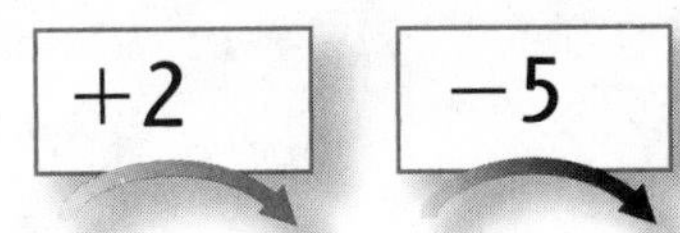

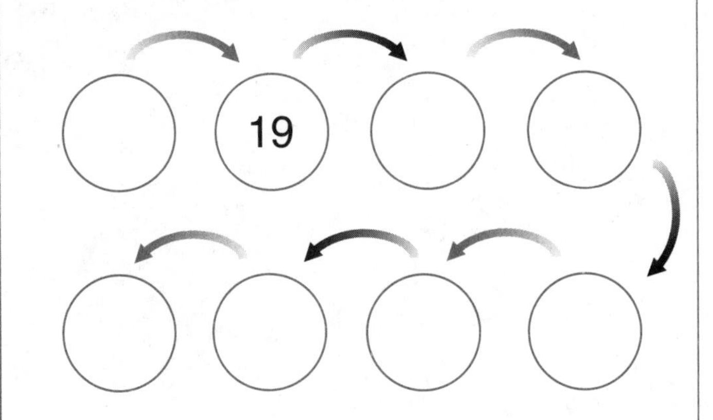

2. Write 6 names for 17.

3. Write the time.

_____ : _____

4. 42 =

_____ tens and _____ ones

86 =

_____ tens and _____ ones

7 =

_____ tens and _____ ones

5. 10 children ordered juice. 13 children ordered milk. How many children ordered drinks?

_____ children

Fill in the diagram; write a number model.

Total	
Part	Part

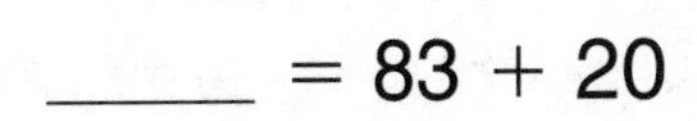

____

6. Solve.

Unit

$50 + 6 =$ _____

_____ $= 83 + 20$

$47 = 77 -$ _____

Date Time

School Supply Store

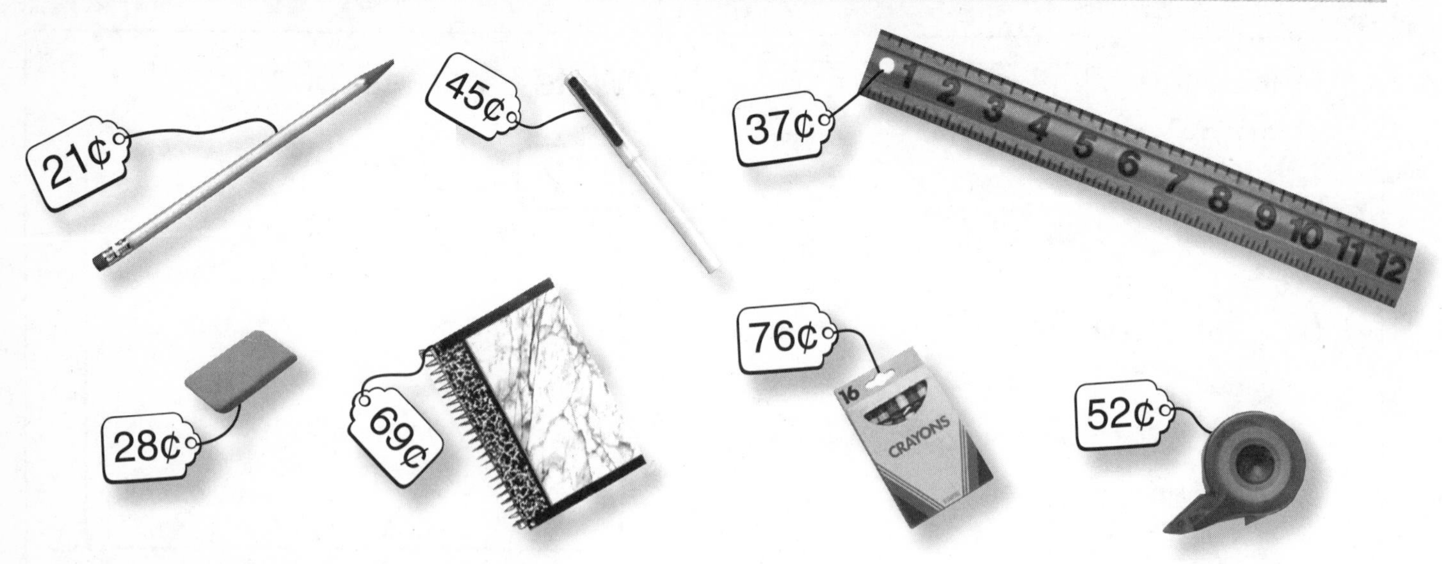

You have $1.00 to spend at the School Store.
Use estimation to answer each question.

Can you buy:	Write *Yes* or *No*.
1. a ruler and a notebook?	______
2. a pen and a ruler?	______
3. a box of crayons and a roll of tape?	______
4. a pencil and a box of crayons?	______
5. 2 rolls of tape?	______
6. a pencil and 2 erasers?	______

7. You want to buy two of the same item.
   List items you could buy two of with $1.00.

   ______________ ______________

   ______________ ______________

8. How many pencils could you buy with $1.00? ______

Date   Time

Comparing Quantities

Reminder: $<$ means *is less than*
$>$ means *is greater than*
$=$ means *is equal to*

Write $<$, $>$, or $=$ for the following problems.

1. 563 _____ 536

49 _____ 80

100 _____ 99

2. 3 nickels _____ 2 dimes

3 dimes _____ 2 quarters

1 dollar _____ 4 quarters

3. 648 _____ 468

$6 + 8$ _____ $8 + 6$

$57 + 38$ _____ $57 + 48$

4. $9 + 6$ _____ $20 - 5$

• • •
• • •   _____ 𝍸 𝍸
• • •

$39 - 25$ _____ $39 - 29$

5. Make up your own problems.

_____ $<$ _____   _____ $<$ _____

_____ $>$ _____   _____ $>$ _____

_____ $=$ _____ $+$ _____   _____ $=$ _____ $+$ _____

Date Time

Math Boxes 4.5

1. What temperature is it?

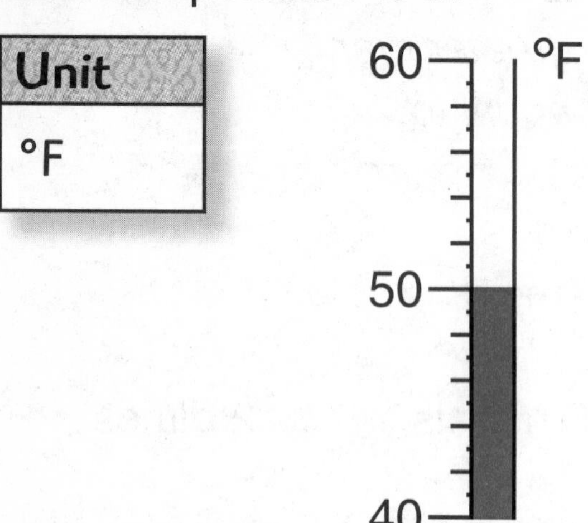

______

2. I bought a sandwich and ice cream. Each cost 35¢. How much did I spend?

______¢

Fill in the diagram and write a number model.

Total	
Part	Part

______

3. Solve.

$$\begin{array}{r} 57 \\ -\ 7 \\ \hline \end{array}$$

______ = 40 + 3

79 − ______ = 70

80 = ______ − 5

4. Fill in the frames.

Rule
− 2

( ) → ( ) → (90) → ( ) → ( )

5. I had \$0.35. I spent \$0.15. How much change do I have?

\$______

6. Fill in the missing numbers.

	112	
121		

Date Time

Shopping

Play *Shopping* with a partner.

Materials

- ❑ *Shopping* cards (*Math Masters,* p. 67)
- ❑ *Math Masters,* p. 61
- ❑ calculator
- ❑ play money for each partner: at least nine $1 bills and eight $10 bills; one $100 bill (optional)

Directions

1. Partners take turns being the "Customer" and the "Clerk."
2. The Customer draws two cards and turns them over. These are the items the Customer is buying.
3. The Customer places the two cards on the parts-and-total diagram—one card in each Part box.
4. The Customer figures out the total cost of the two items without using a calculator.
5. The Customer counts out bills equal to the total cost. The Customer places the bills in the Total box on the parts-and-total diagram.
6. The Clerk uses a calculator to check that the Customer has figured the correct total cost.
7. Partners switch roles.
8. Play continues until all eight cards have been used.

Another Way to Play

Instead of counting out bills, the Customer says or writes the total. The Customer gives the Clerk a $100 bill to pay for the items. The Clerk must return the correct change.

Date Time

Shopping Problems

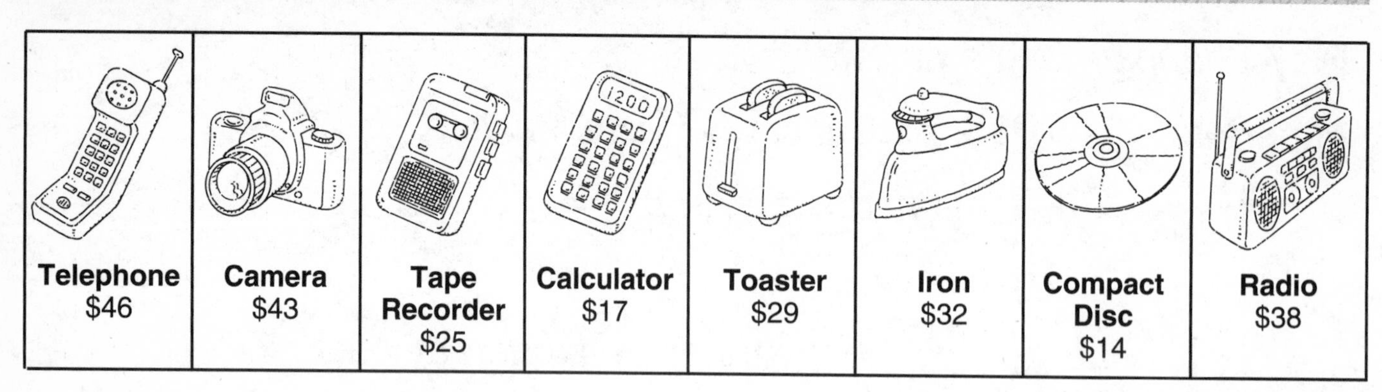

Total	
Part	Part

1. You buy a telephone and an iron.

   What is the total cost? $______

   Number model: ____________________

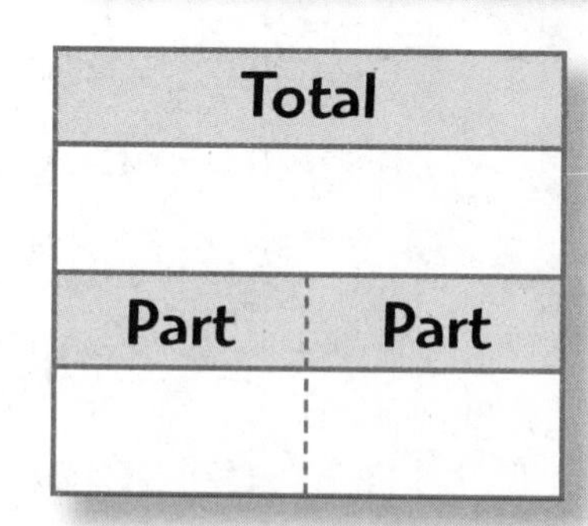

2. You buy a radio and a calculator.

   What is the total cost? $______

   Number model: ____________________

Solve each problem.

3. You bought two items. The total cost is exactly $60. What did you buy?

   ____

4. You bought two items. Together they cost the same as a telephone. What did you buy?

   ____

5. You bought three items. The total cost is exactly $60. What did you buy?

   ____

Date Time

Math Boxes 4.6

1. A.M. temperature was 50°F.
   P.M. temperature is 68°F.

   What was the change? _____ °F

   Fill in the diagram and write the number model.

Start	Change	End

   ____

2. I have \$1.00. Can I buy two 45¢ ice cream bars?

   ____

3. Continue.

   258, 248, _____, _____,

   _____, _____, _____,

   _____, _____, _____,

   _____, _____, _____

4. Circle names that belong.

   \$1.00

10 dimes	4 quarters
18 nickels	100 pennies
5 dimes	5 nickels

5. What temperature is it?

   ____

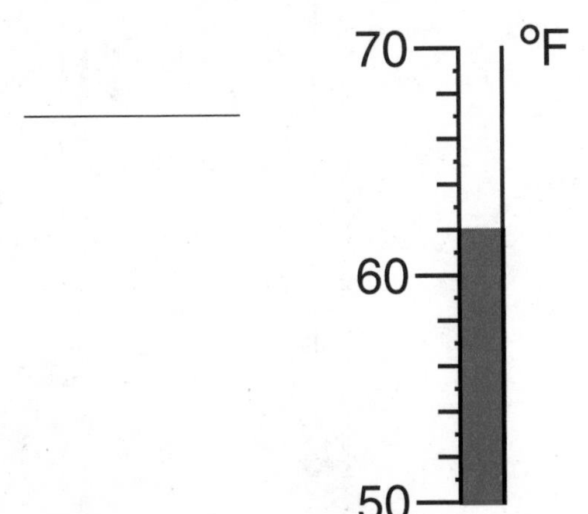

6. Use >, <, or =.

   20 + 30 _____ 30 + 33

   716 _____ 617

   109 − 9 _____ 100

Date Time

Measuring Lengths with a Tape Measure

1. Measure the height from the top of your desk or table to the floor. Measure to the nearest inch.

The height from my desk or table to the floor is about _____ inches.

---

2. Measure the height from the top of your chair to the floor. Measure to the nearest inch.

The height from the top of my chair to the floor is about _____ inches.

---

3. Measure the width of your classroom door.

The classroom door is about _____ inches wide.

Date Time

Measuring Lengths with a Tape Measure (cont.)

4. Open your journal so that it looks like the drawing below.

a. Measure the long side and the short side to the nearest inch.

The long side is about _____ inches.

The short side is about _____ inches.

b. Now measure your journal to the nearest centimeter.

The long side is about _____ centimeters.

The short side is about _____ centimeters.

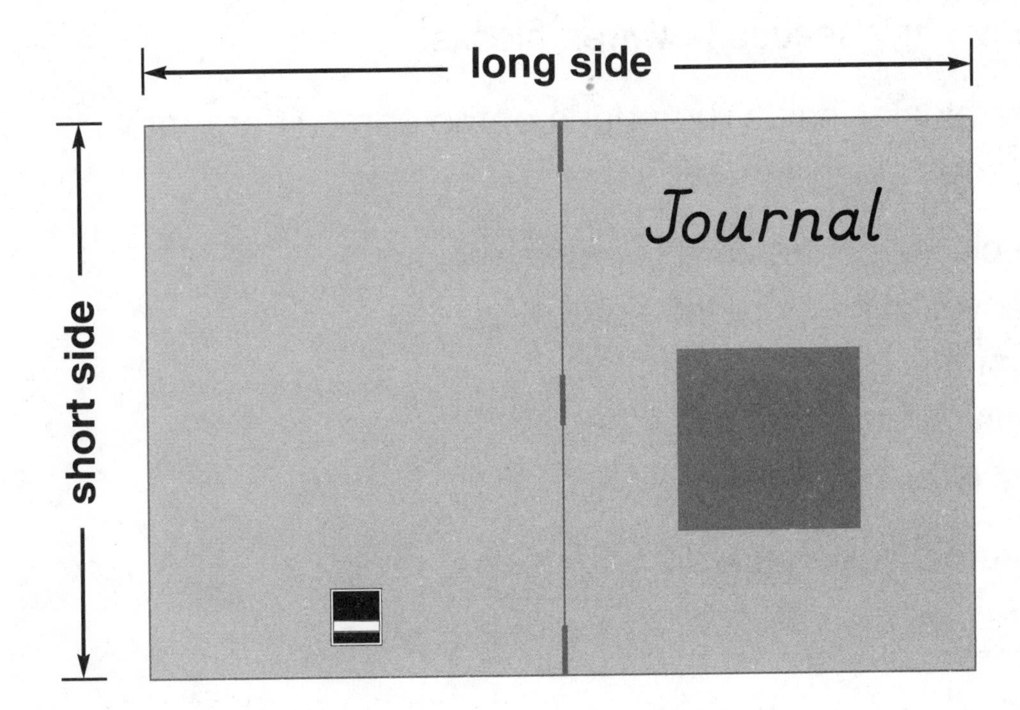

Date Time

Tiling Surfaces with Shapes

Materials
- ❑ pattern blocks
- ❑ slates
- ❑ sheets of paper
- ❑ Pattern-Block Template
- ❑ scissors
- ❑ Everything Math Deck cards, if available

1. Pick one pattern-block shape. **Tile** a card by covering it with blocks of this shape.

   - Lay the blocks flat on the card.
   - Don't leave any spaces between blocks.
   - Keep the blocks inside the edges of the card. There may be open spaces along the edges.

   Count the blocks on the card. If a space could be covered by more than half of a block, count the space as one block. Do not count spaces that could be covered by less than half of a block.

   Which pattern-block shape did you use?

   ____

   Number of blocks needed to tile the card:

   ____

   Trace the card. Use your Pattern-Block Template to draw the blocks you used to tile the card.

Date Time

Tiling Surfaces with Shapes (cont.)

2. Use Everything Math Deck cards to tile both a slate and a Pattern-Block Template. How many cards were needed to tile them?

Slate: __________ cards

Pattern-Block Template: __________ cards

3. Fold a sheet of paper into fourths. Cut the fourths apart. Use them to tile larger surfaces, such as a desktop.

Surface	Number of Fourths
____________________	____________________
____________________	____________________

Follow-Up

With a partner, find things in the classroom that are tiled or covered with patterns. Make a list. Be ready to share your findings.

An Attribute Rule

Choose an Attribute Rule Card. Copy the rule below.

Rule: ______________________________

Draw or describe all of the attribute blocks that fit the rule.

Draw or describe all of the attribute blocks that *do not* fit the rule.

These blocks fit the rule:

____

____

____

____

These blocks *do not* fit the rule:

____

____

____

____

Date Time

Math Boxes 4.7

1. Solve.

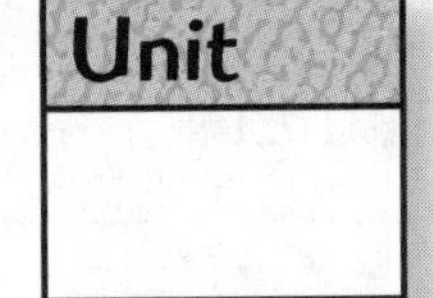

$5 + 3 =$ _____

$50 + 30 =$ _____

$\begin{array}{r} 6 \\ +\ 3 \\ \hline \end{array}$ $\begin{array}{r} 60 \\ +\ 30 \\ \hline \end{array}$

2. I can bring 70 pounds of luggage onto an airplane. I have an 18-pound box and a 59-pound suitcase. Can I bring both? Estimate.

_____

3. 20 airplanes. 8 take off.

How many left? _____ airplanes
Fill in the diagram and write a number model.

____

4. Show 8:50 P.M.

5. Fill in the frames.

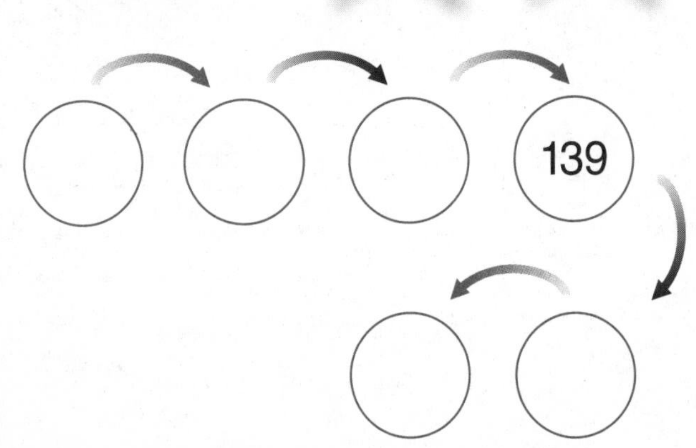

6. Write the fact family.

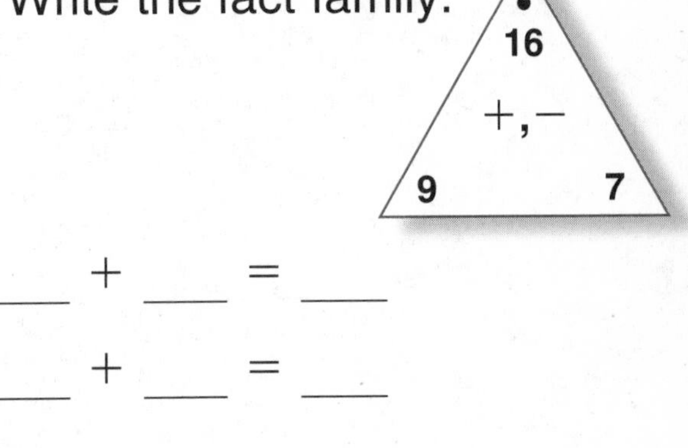

___ + ___ = ___

___ + ___ = ___

___ − ___ = ___

___ − ___ = ___

Date Time

Addition Practice

Add. Show your work in the workspaces. Check your work. Write a number model to show the ballpark estimate.

1. $\begin{array}{r} 39 \\ + 26 \\ \hline \end{array}$ Answer: ____ Ballpark estimate: ____	**2.** 18 + 45 Answer: ____ Ballpark estimate: ____	**3.** 52 + 28 Answer: ____ Ballpark estimate: ____
4. 54 + 79 Answer: ____ Ballpark estimate: ____	**5.** 115 + 32 Answer: ____ Ballpark estimate: ____	**6.** $\begin{array}{r} 327 \\ + 146 \\ \hline \end{array}$ Answer: ____ Ballpark estimate: ____

Add. In each problem, use the first sum to help you find the other two sums.

7. 17 + 8 = ______

17 + 8 + 25 = ______

17 + 8 + 25 + 12 = ______

8. $\begin{array}{r} 15 \\ + 9 \\ \hline \end{array}$ $\begin{array}{r} 15 \\ 9 \\ + 6 \\ \hline \end{array}$ $\begin{array}{r} 15 \\ 9 \\ 6 \\ + 22 \\ \hline \end{array}$

9. 19 + 6 = ______

19 + 6 + 5 = ______

19 + 6 + 5 + 70 = ______

10. $\begin{array}{r} 24 \\ + 4 \\ \hline \end{array}$ $\begin{array}{r} 24 \\ 4 \\ + 7 \\ \hline \end{array}$ $\begin{array}{r} 24 \\ 4 \\ 7 \\ + 35 \\ \hline \end{array}$

Date Time

"What's My Rule?" and Frames and Arrows

Find the rule. Then write the missing numbers in each table.

Unit
°F

1.

in	out
30	37
25	
	43
8	
	19

2.

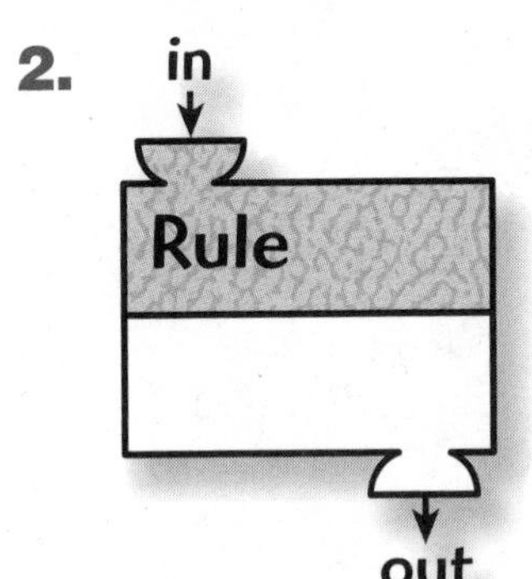

in	out
50	30
	50
10	
20	
	42

Fill in the frames.

3.

+5

+10

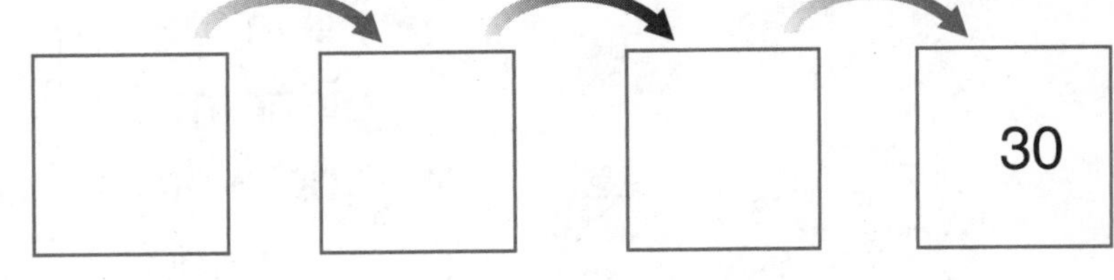

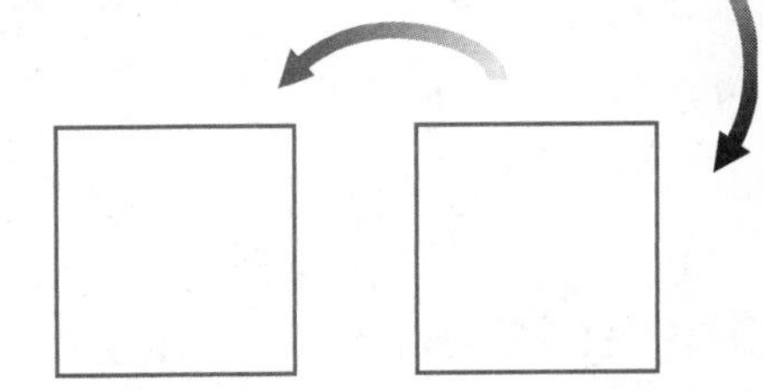

4.

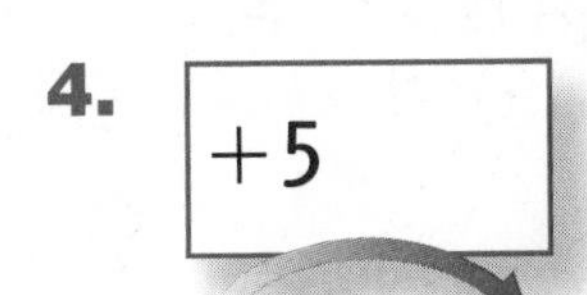

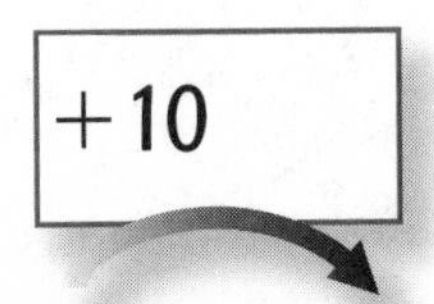

	25	35	

Date Time

Math Boxes 4.8

1. Measure the length of this line.

____

about ________ cm

about ________ in.

2. I have $1.00. Can I buy a 28¢ pen and a 52¢ marker?

____

3. How much?

$________

4. Jerry has 9 action figures. His twin has 1 fewer. How many in all? ______ action figures

Fill in the diagram and write a number model.

Total	
Part	Part

____

5. Fill in the rule and the missing numbers.

Rule

in	out
$1.13	$2.13
$1.63	$2.63
	$5.29
$6.46	

6. I bought a radio for $67.00.
I paid with $100.00.
How much change did I get?

____

Date Time

Addition Practice

Add. Show your work. Check your work. Write a number model to show your ballpark estimate.

1. 59 + 8 Answer: ______ Ballpark estimate: ______	2. 67 + 7 Answer: ______ Ballpark estimate: ______	3. 47 + 32 Answer: ______ Ballpark estimate: ______
4. 43 + 37 Answer: ______ Ballpark estimate: ______	5. 28 + 57 Answer: ______ Ballpark estimate: ______	6. 49 + 29 Answer: ______ Ballpark estimate: ______
7. 58 + 26 Answer: ______ Ballpark estimate: ______	8. 122 + 53 Answer: ______ Ballpark estimate: ______	9. 136 + 157 Answer: ______ Ballpark estimate: ______

Date Time

The Time of Day

For Problems 1–4, draw the hour hand and the minute hand to show the time.

1. Dina got up at 7:00.

She had breakfast an hour later.

Show the time when she had breakfast.

2. Mel left home at 8:15.

It took him half an hour to get to school. Show the time when he arrived at school.

3. Mia finished reading a story at 10:30.

It took her 15 minutes. Show the time when she started reading.

4. The second graders went on a field trip.

They left school at 12:30.

They got back 2 hours later.

Show the time when they got back.

5. The clock shows when Bob went to bed.

He went to sleep 15 minutes later.

At what time did he go to sleep?

____ : ____

Date Time

Math Boxes 4.9

1. Solve.

Unit

23 + 30 = ____

86 = ____ + 50

____ = 40 + 59

67 + 30 = ____

2. How much?

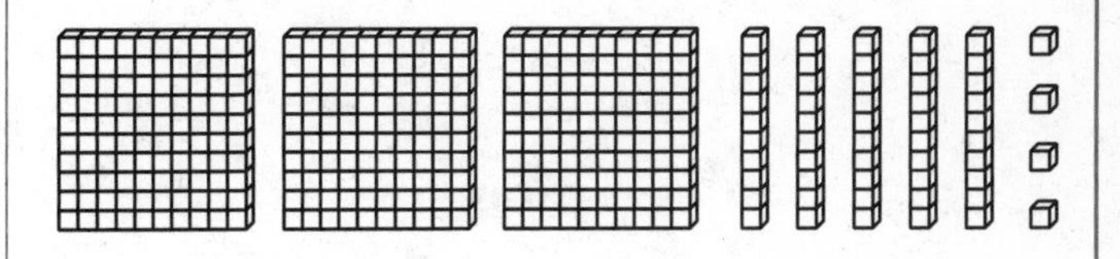

____

3. Write the number. Be careful!

6 tens

3 ones

8 hundreds

____

4. Draw a line segment 6 cm long. Underneath it, draw a line segment that is 2 cm longer.

5. Bought a model dinosaur for 37¢. Paid with 2 quarters. How much change?

____

6. I had 17 tulips. I planted 20 more. How many do I have now? ____ tulips

Fill in the diagram and write a number model.

____

Date Time

Math Boxes 4.10

1. What time is it?

_____ : _____

What time will it be in 20 minutes?

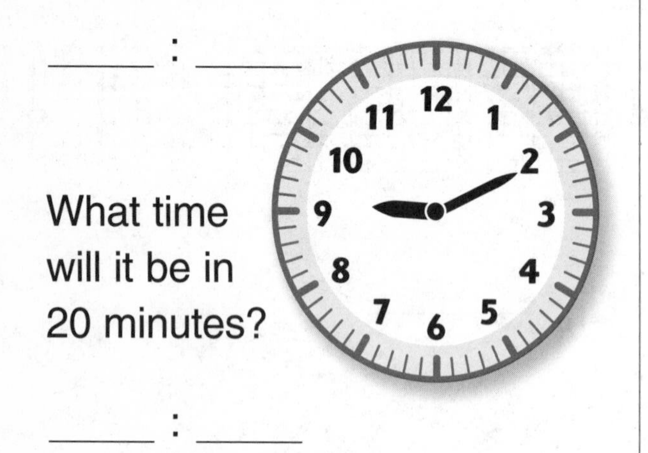

_____ : _____

2. What is the temperature?

____

Would you wear a coat?

____

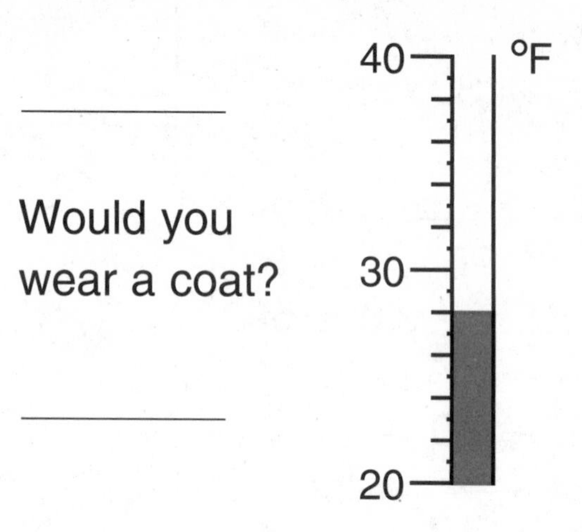

3. Use the digits 5, 7, and 3. Write the smallest possible number. Use each digit once.

____

4. Solve.

Unit

18 − 9 = _____

180 − 90 = _____

1,800 − 900 = _____

5. Peanuts cost 38¢ at the circus. Popcorn costs 57¢. Will $1.00 pay for both?

____

6. Put an X on the digit in the tens place in each number.

3 6 2 1, 0 4 3

1, 2 0 9 5 9 6

Date Time

Math Boxes 5.1

1. Solve.

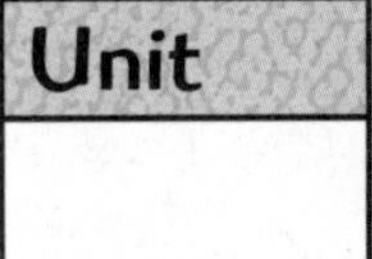

$$\begin{array}{r} 16 \\ -\ 8 \\ \hline \end{array} \qquad \begin{array}{r} 14 \\ -\ 8 \\ \hline \end{array}$$

2. Show 7:20.

3. The temperature was 73°F in the afternoon. It dropped to 43°F in the evening. What was the difference?

_____ °F Fill in the diagram and write the number model.

Start	Change	End

____

4. Write 6 names for 20.

20

5. Ronald found 1 quarter and 5 dimes in a coat pocket. He found 3 nickels and 4 pennies in a pants pocket. How much money did he find in all?

_____

6. Make ballpark estimates. Write a number model for your estimates.

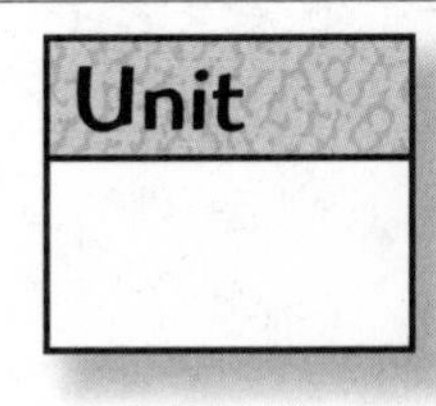

28 + 67

_____ + _____ = _____

51 + 38

_____ + _____ = _____

Date Time

Attributes

Materials

- ❑ attribute blocks
- ❑ sheet of paper
- ❑ red, yellow, and blue crayons or pencils

Solve each problem. On a separate sheet of paper, trace and color the blocks to show your answers.

1. Find 2 blocks that are NOT the same size, NOT the same shape, and NOT the same color.

2. Find 2 blocks that have the same shape, but are NOT the same size and NOT the same color.

3. Find 3 blocks that are the same size and the same color, but are NOT the same shape.

4. Find 4 small blocks that are the same color, but are NOT the same shape.

Date Time

Math Boxes 5.2

1. What time is it?

_____ : _____

What time will it be in 15 minutes?

_____ : _____

2. Solve.

Unit

3 + 5 = _______

30 + 50 = _______

300 + 500 = _______

_______ = 6 + 8

_______ = 60 + 80

_______ = 600 + 800

3. Spend 68¢. Pay with $1.00. How much change?

_____

4. 25 books. Bought 15 more.

How many now? _____ books

Fill in the diagram and write a number model.

Start	Change	End

____

5. Find the rule. Complete the table.

Rule

in	out
11	6
19	
15	10
	8
	12

6. Count by quarters to $2.00.

$0.25, _______, _______,

_______, _______, _______,

_______, _______

Date Time

Polygons

Triangles

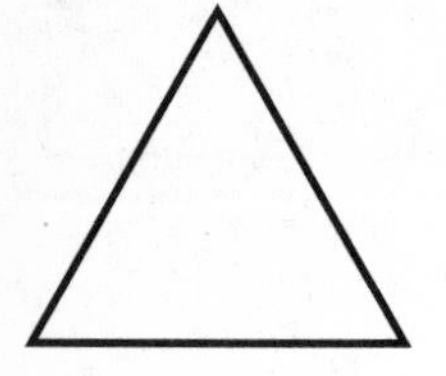
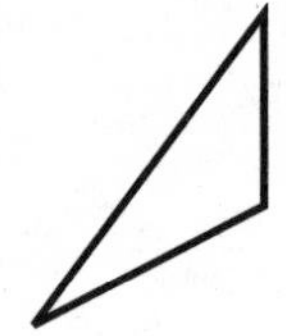
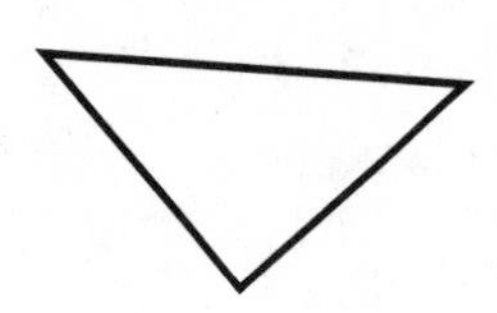

Quadrangles or Quadrilaterals

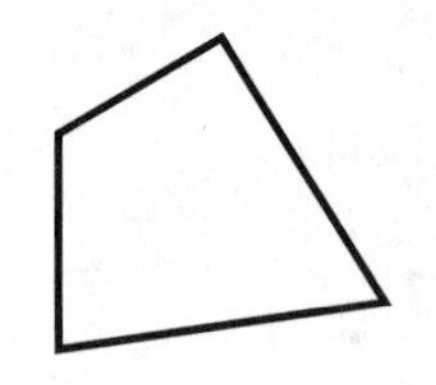
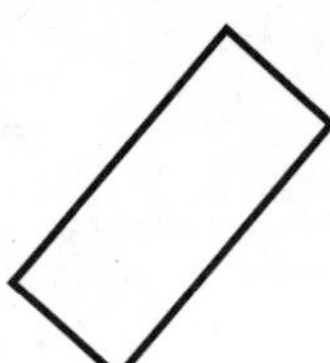
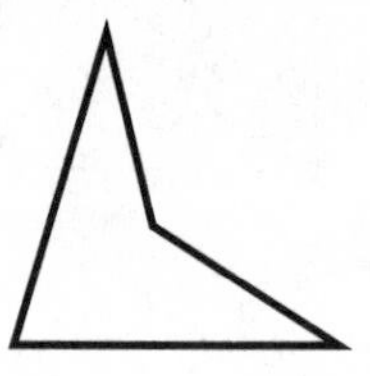

Pentagons

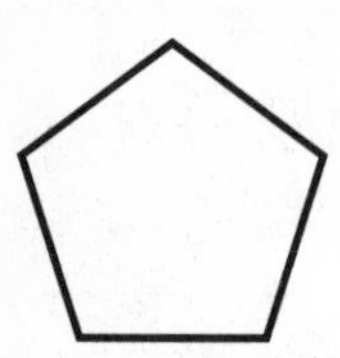

Hexagons

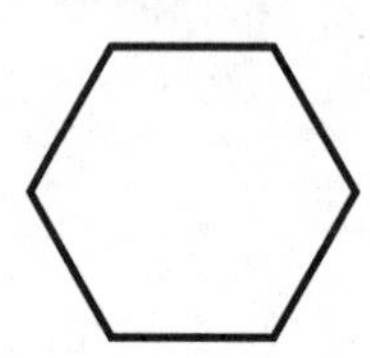
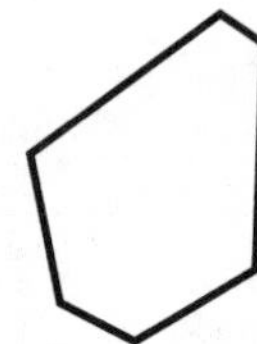
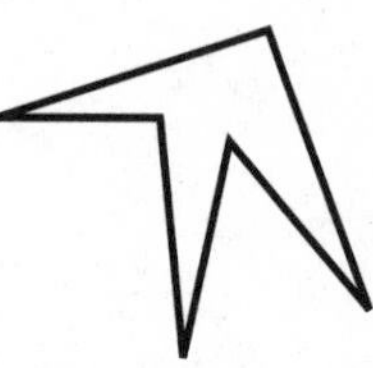

Heptagons

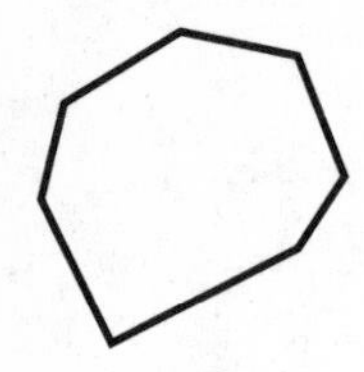
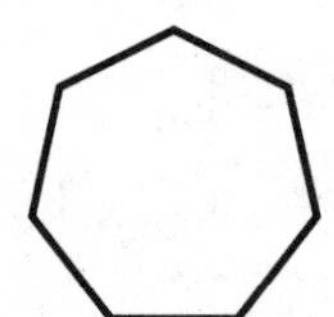

Octagons

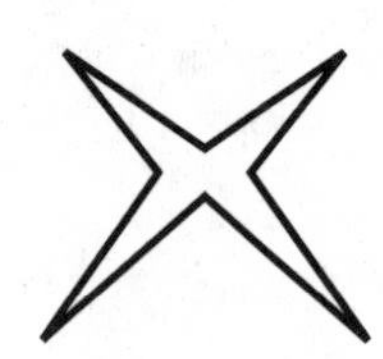
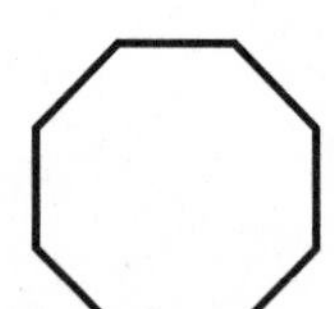
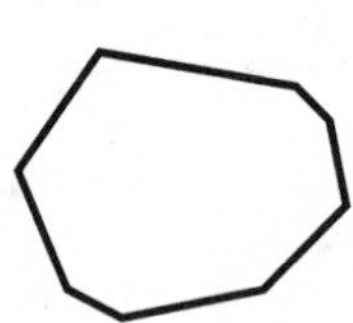

These are NOT polygons.

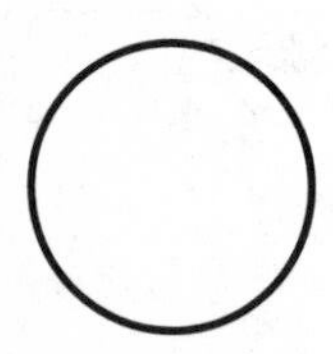
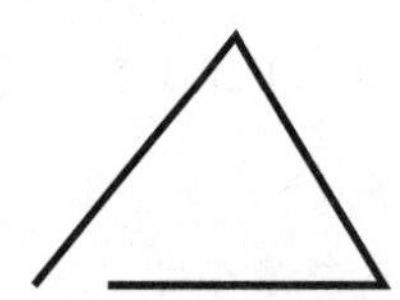
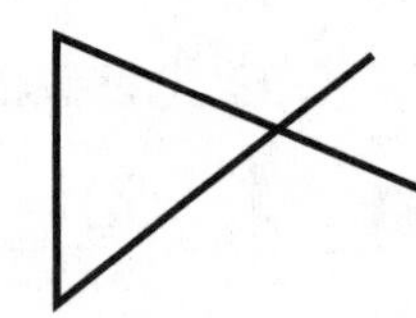
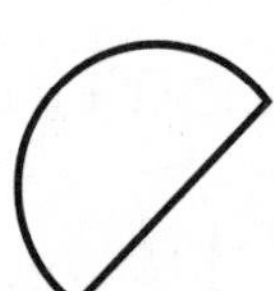

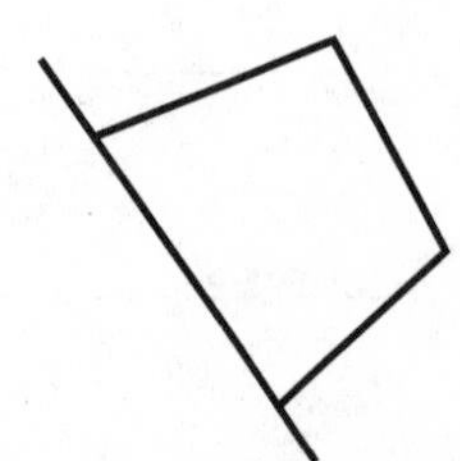
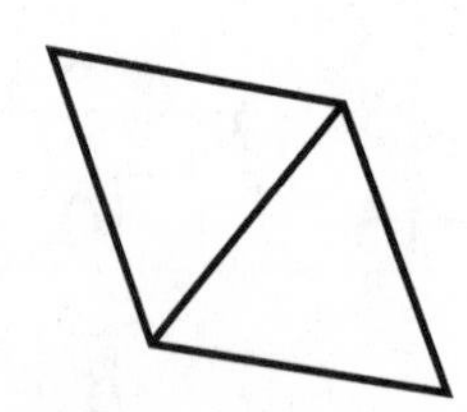
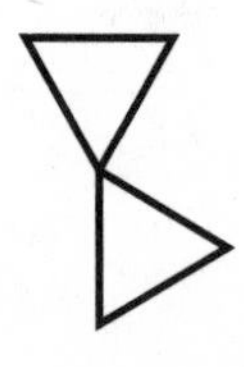

Date Time

Math Boxes 5.3

1. The temperature was 73°F. It got 13°F colder. What is the temperature now? ______ °F
Fill in the diagram and write a number model.

Start	Change	End

____

2. Solve.

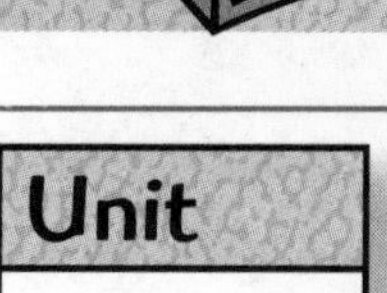

9 − 5 = ______

______ = 90 − 50

900 − 500 = ______

______ = 9,000 − 5,000

3. Fill in the diagram and write a number model.

Total	
Part	Part
12	18

____

4. Find the rules.

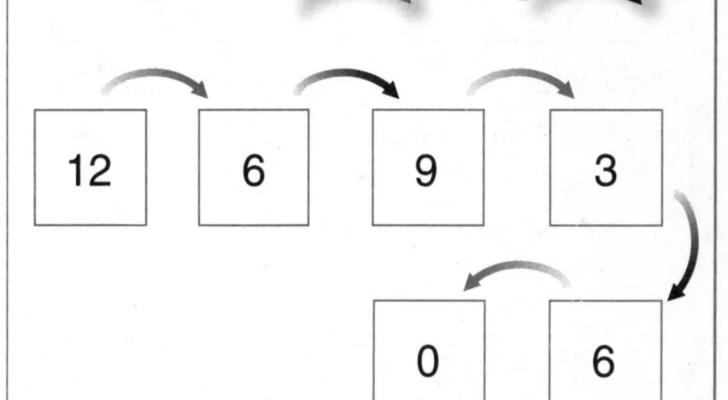

5. Use the partial-sums algorithm to solve. Show your work.

$$\begin{array}{r} 45 \\ +\ 36 \\ \hline \end{array}$$

6. What number?

______

Date Time

Using Secret Codes to Draw Shapes

1. The codes show how to connect the points with line segments. Can you figure out how each code works? Talk it over with your partner.

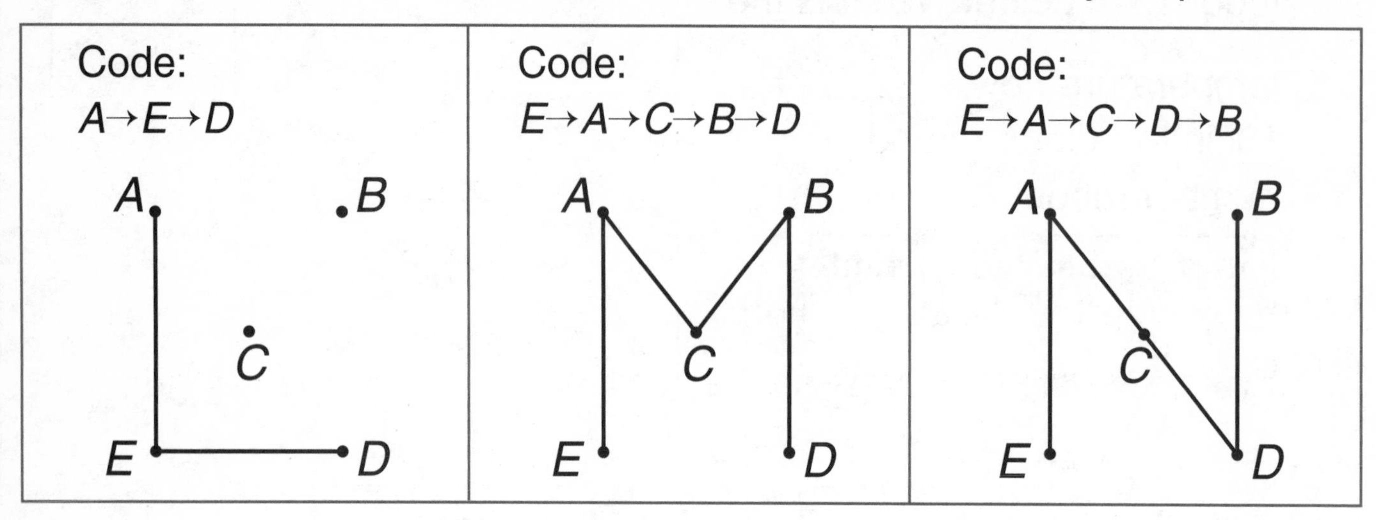

2. Use each code. Draw line segments using your straightedge.

Code: A→B→C→A	Code: A→B→D→E→A	Code: A→B→C→E→D→C→A
A. .B C E• •D	A. .B C E• •D	A. .B C E• •D
Code: B→A→C→E→D→B	Code: A→B→C→E→A	Code: C→D→E→C→B→A→C
A. .B C E• •D	A. .B C E• •D	A. .B C E• •D

Date Time

Math Boxes 5.4

1. Solve.

Unit

15 − 8 = _____

_____ = 35 − 8

65 − 8 = _____

95 − 8 = _____

_____ = 55 − 8

2. Joe has $2.00 and spends 65¢. How much money is left?

____

3. Complete the table.

Rule
Add 20

in	out
36	
52	
	63
	49

4. Find the differences.

32°F and 53°F _______

37°C and 19°C _______

75°F and 93°F _______

5. Use your Pattern-Block Template. Draw a trapezoid.

How many sides? _____ sides

6. Solve.

Unit

15 + _____ = 100

100 = 35 + _____

25 + _____ = 100

80 + _____ = 100

Date Time

Parallel or Not Parallel?

These line segments are **parallel.**

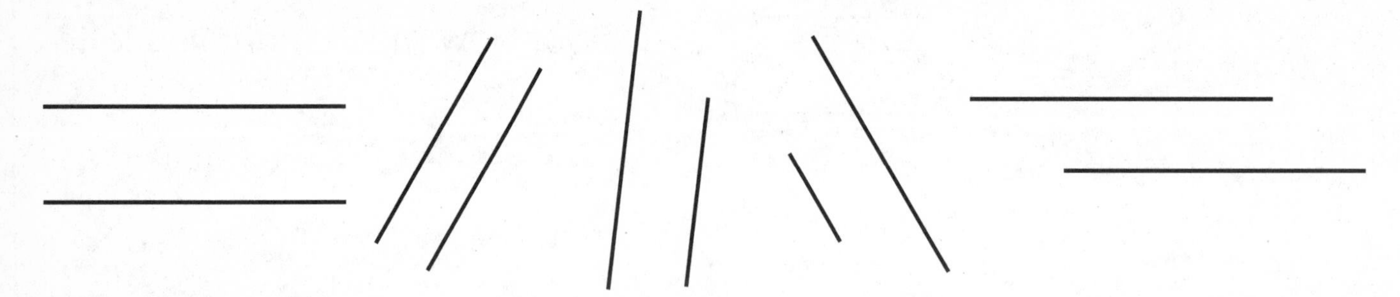

These line segments are **not parallel.**

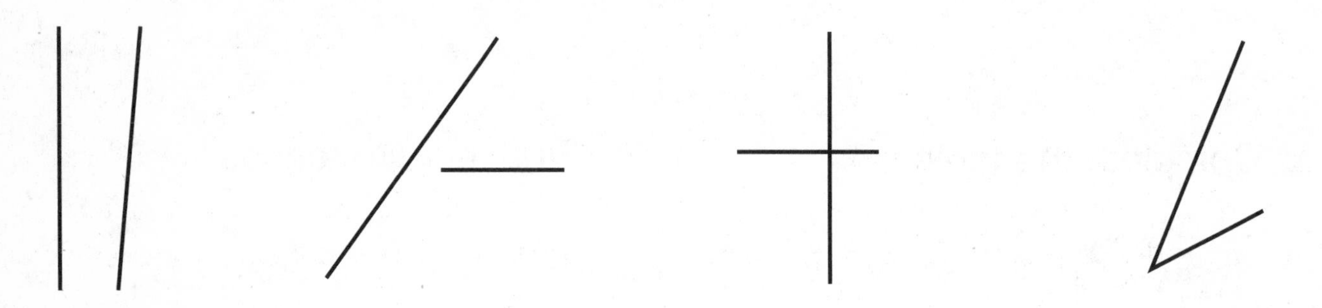

Quadrangles (Quadrilaterals)

These polygons are all quadrangles (quadrilaterals).

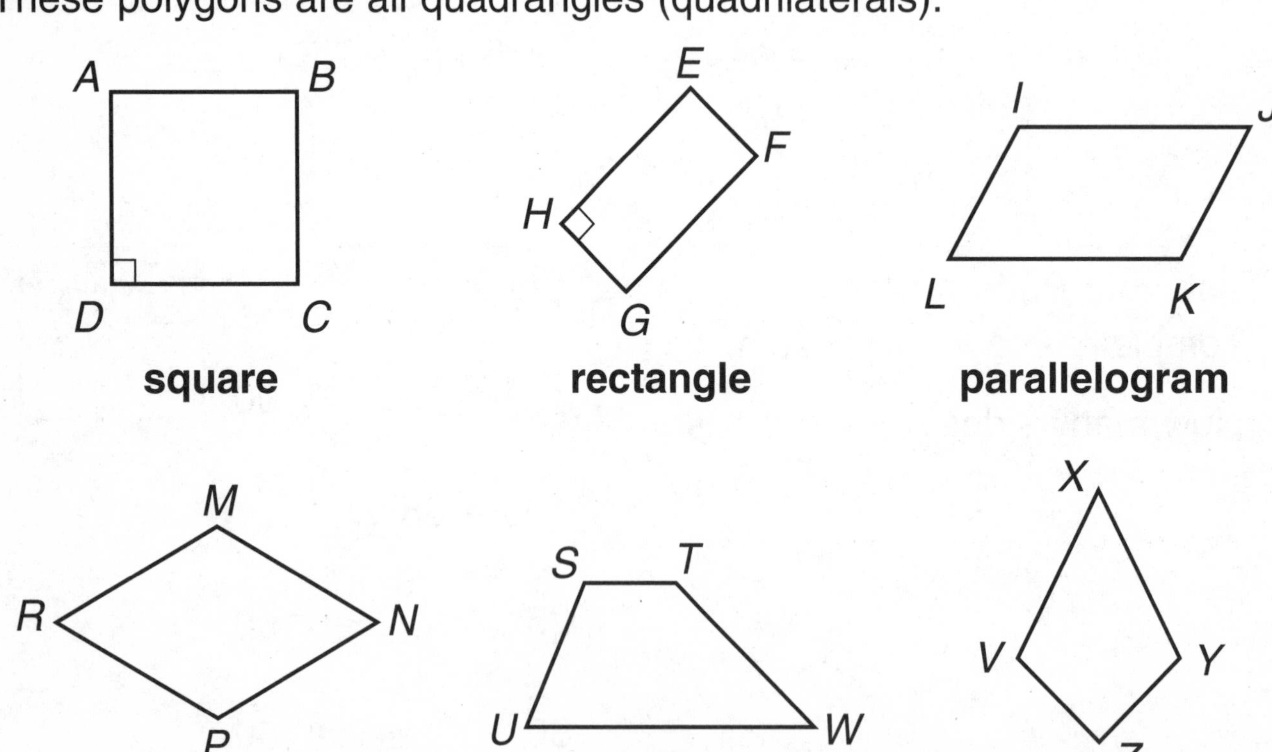

Date Time

Math Boxes 5.5

1. Solve.

Unit

$9 + 4 =$ _____

_____ $= 19 + 4$

_____ $= 39 + 4$

_____ $= 59 + 4$

$89 + 4 =$ _____

2. Count by quarters to \$3.00. Start at \$1.00.

\$1.00, _______, _______,

_______, _______, _______,

_______, _______, _______

3. Use your ruler. Draw $\overline{MS}$.

$\overline{MS}$ is _____ cm long.

4. Does the arrow point to a vertex or a side?

____

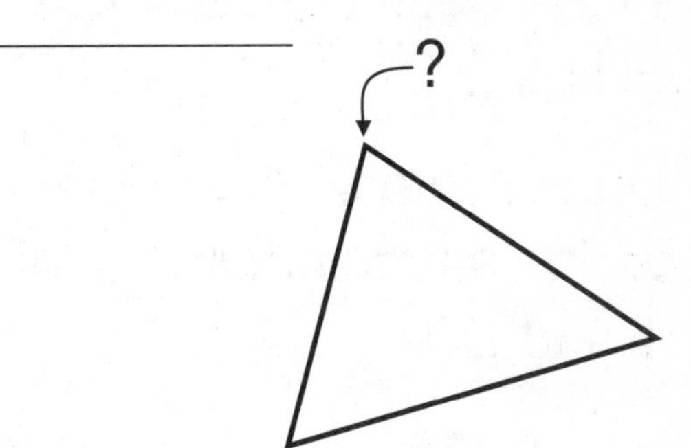

5. Use your Pattern-Block Template. Draw a rhombus.

There are _____ sides.

6. 15 goldfish. 18 angelfish.

How many in all? _____ fish

Fill in the diagram and write a number model.

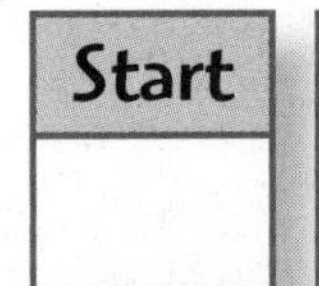

____

Date Time

Parallel Line Segments

Use a straightedge.

1. Draw line segments *AB* and *CD*.

Are line segments *AB* and *CD* parallel?

____

2. Draw line segment *EF*.

Are line segments *AB* and *EF* parallel?

____

A B

C D

E F

---

3. Draw line segment *LM*.

4. Draw a line segment that is parallel to line segment *LM*. Label its endpoints *R* and *S*.

5. Draw a line segment that is NOT parallel to line segment *LM*. Label its endpoints *T* and *U*.

L

M

---

6. Draw a quadrangle that has NO parallel sides.

Use with Lesson 5.5.

Date Time

Parallel Line Segments (cont.)

7. Draw a quadrangle in which opposite sides are parallel.

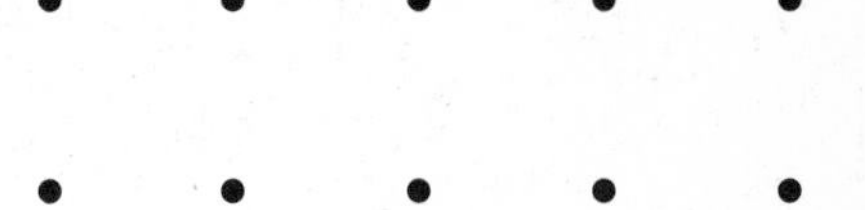

8. Draw a quadrangle in which all four sides are the same length.

What is another name for this shape?

____

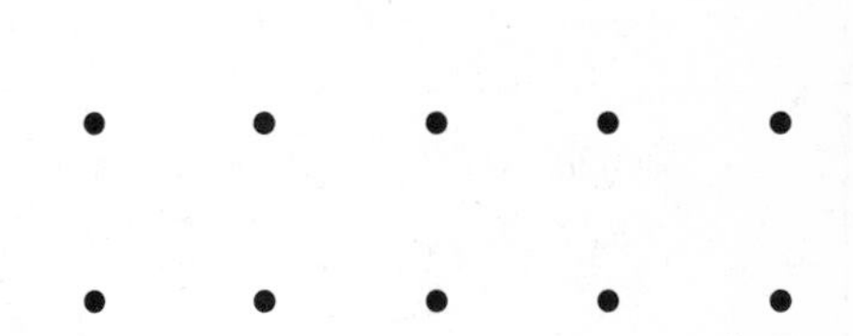

9. Draw a quadrangle in which 2 opposite sides are parallel and the other 2 opposite sides are NOT parallel.

Math Boxes 5.6

1. Find the rules.

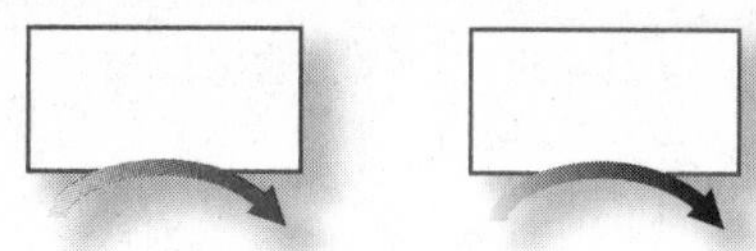

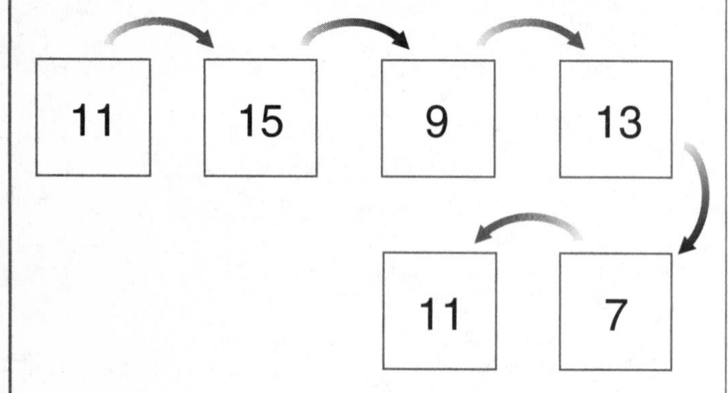

2. 32 second graders. 5 moved. How many now?

_____ second graders

Fill in the diagram and write a number model.

Start	Change	End

____

3. Circle the parallel lines.

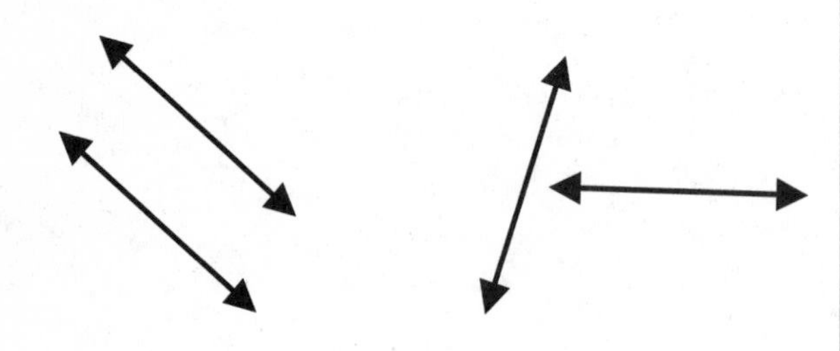

4. Use your Pattern-Block Template. Draw a hexagon.

There are _____ sides.

5. Complete the number grid.

442		

6. Solve.

Unit

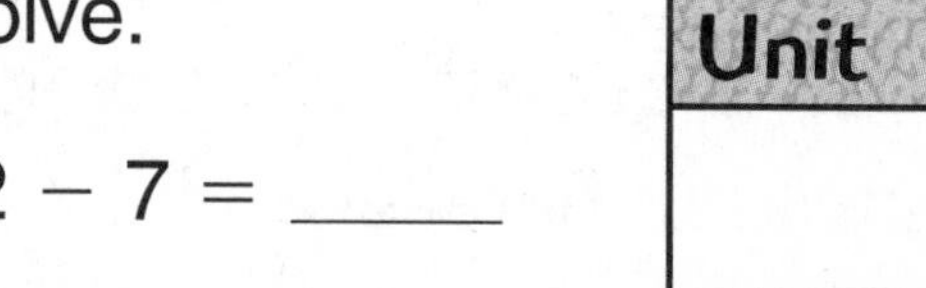

$12 - 7 =$ _____

$32 - 7 =$ _____

_____ $= 52 - 7$

_____ $= 92 - 7$

$62 - 7 =$ _____

Date Time

Math Boxes 5.7

1. Make ballpark estimates. Write a number model for each estimate.

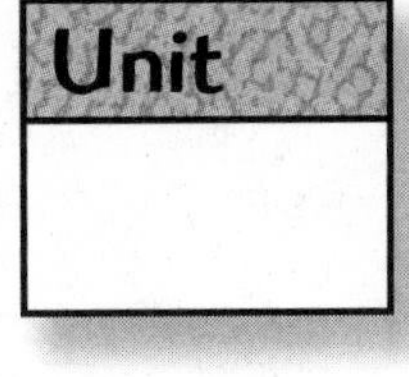

32 + 59

_____ + _____ = _____

51 + 27

_____ + _____ = _____

2. Write 6 names in the 32-box.

32

3. Each cookie costs 30¢. I have $1.00. Can I buy 3 cookies?

_____

4. Use the partial-sums algorithm to solve. Show your work.

$$\begin{array}{r} 39 \\ +\ 46 \\ \hline \end{array}$$

5. This is a trapezoid. Put an **X** on the line segments that are parallel.

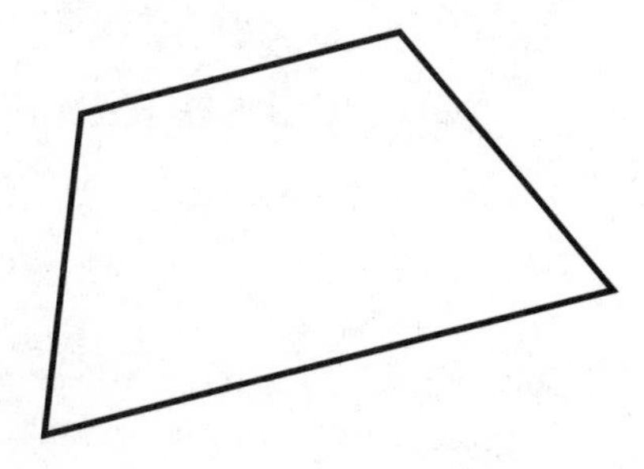

6. Use 5, 7, and 12. Write 2 addition and 2 subtraction facts.

_____

_____

_____

_____

Date Time

3-D Shapes Poster

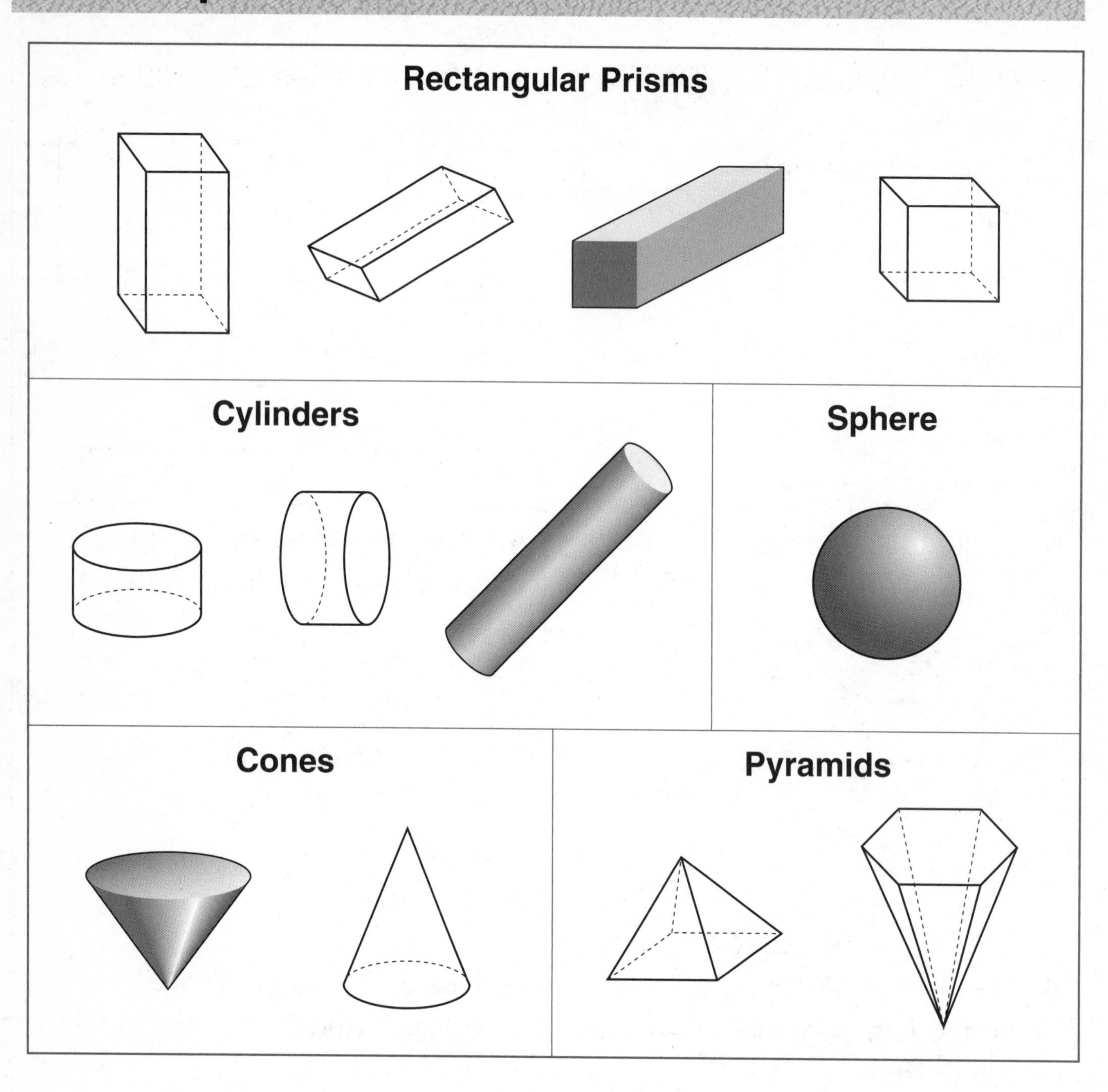

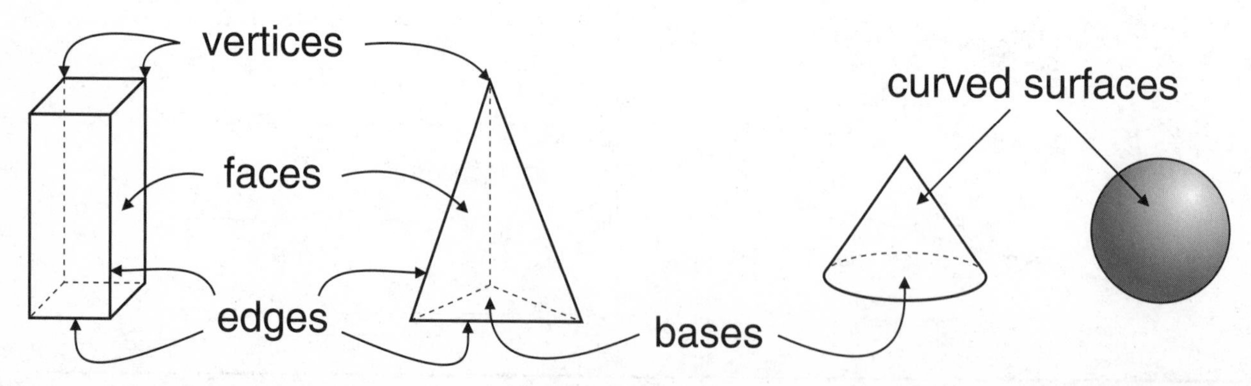

Use with Lesson 5.7.

Date Time

What's the Shape?

Write the name of each shape.

1.

____

2.

____

3.

____

4.

____

5.

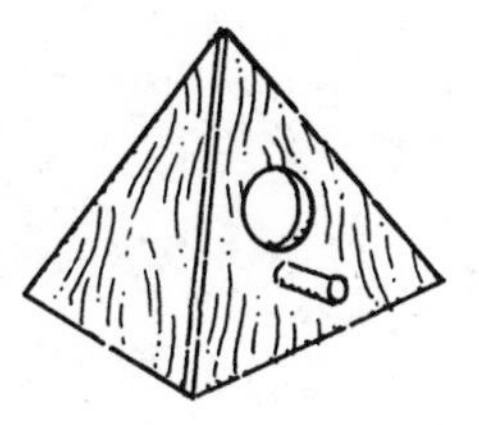

____

6.

____

7.

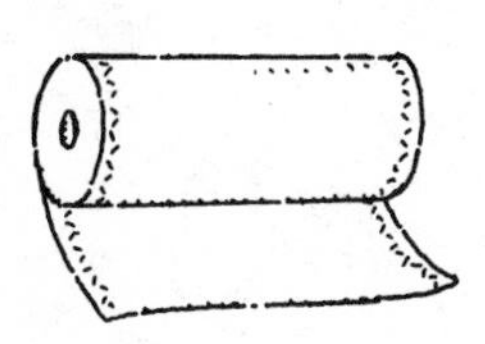

____

8.

____

9.

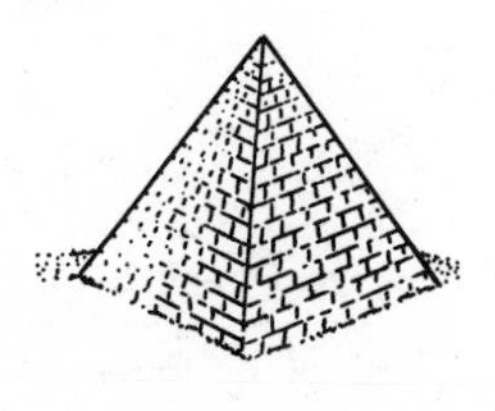

____

10.

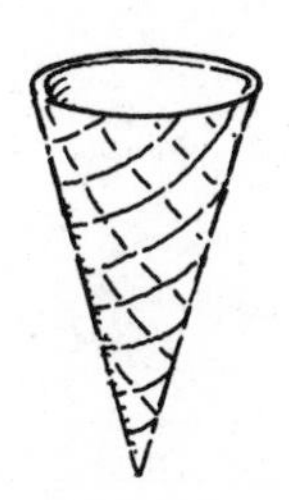

____

Date Time

Connecting Points

Draw a line segment between each pair of points.
Record how many line segments you drew.

Example

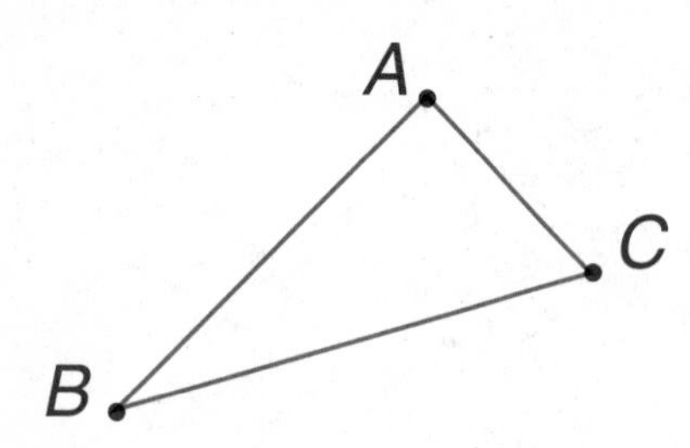

3 points

3 line segments

1.

P

A

L

U

4 points

_____ line segments

2.

R

O

E

S

I

5 points

_____ line segments

3. Connect the points in order from 1 to 3. Use a straightedge.

Find 3 triangles.

Try to find the fourth triangle.

Color a 4-sided figure.

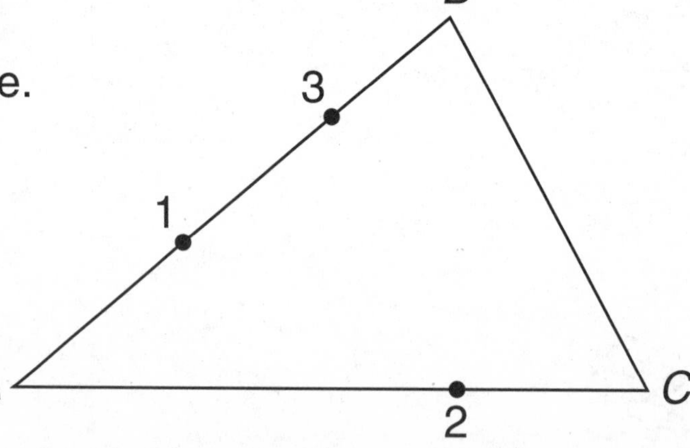

Date Time

Math Boxes 5.8

1. Draw a quadrangle.
Make 2 sides parallel.

2. What is the temperature?

______

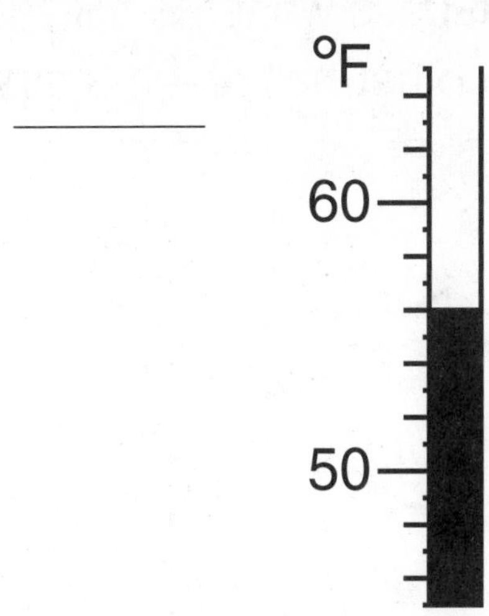

3. Write the names of 3 objects shaped like cylinders.

____

____

____

4. Match quarters to ¢.

5 quarters	150¢
6 quarters	250¢
7 quarters	125¢
10 quarters	175¢

5. 639 has

_____ hundreds

_____ tens

_____ ones

6. 1 dozen eggs. 3 cracked.

How many left? _____ eggs
Fill in the diagram and write a number model.

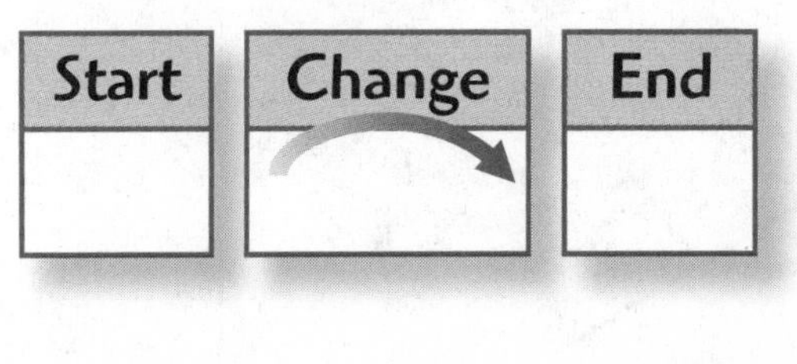

____

Date Time

Symmetrical Shapes

Each picture below shows half of a shape on your Pattern-Block Template. Guess what the full shape is. Then use your template to draw the other half of the shape. Write the name of the shape.

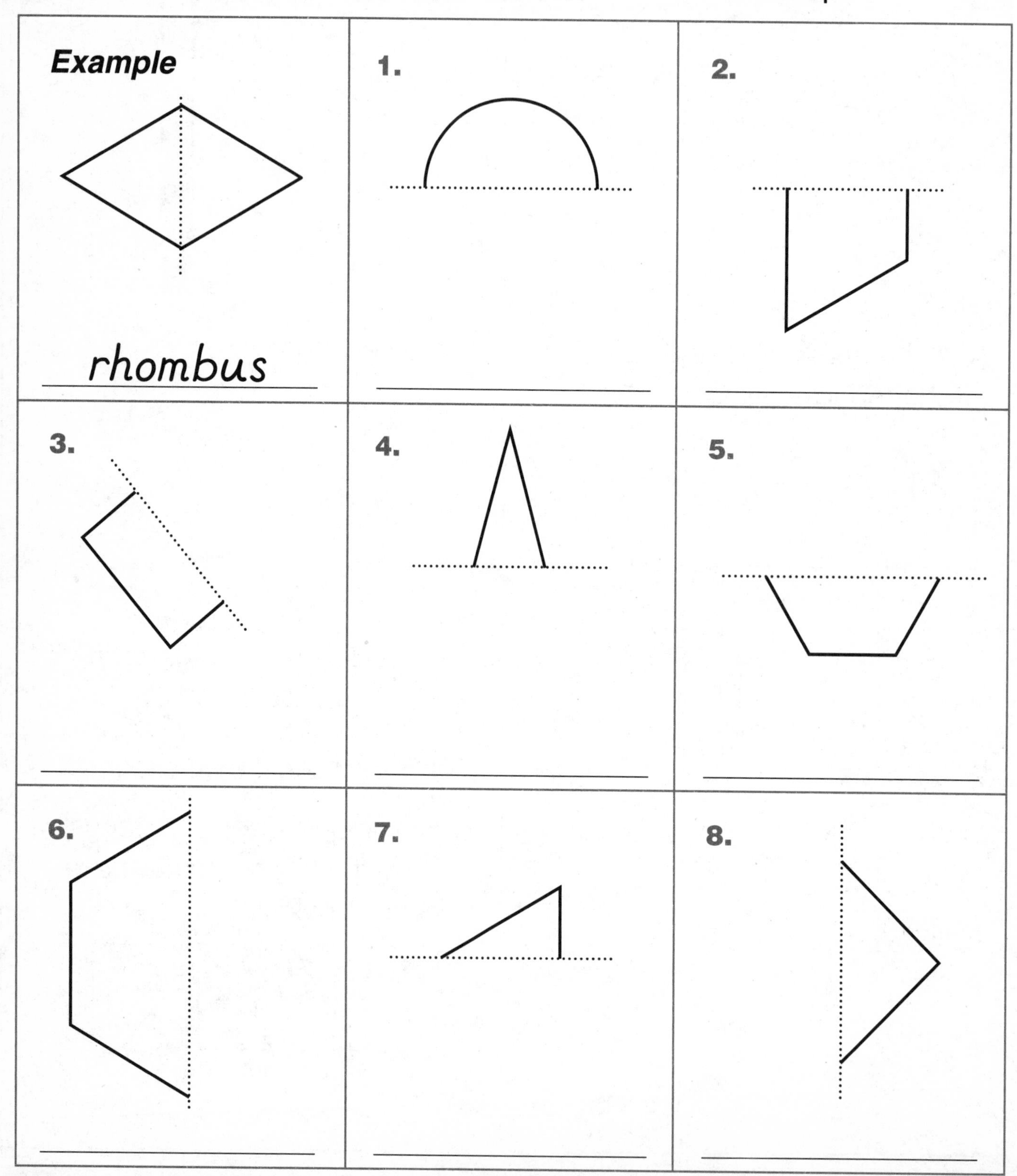

Use with Lesson 5.9.

Date Time

Math Boxes 5.9

1. What time is it?

_____ : _____

In $\frac{1}{2}$ hour,
it will be

_____ : _____.

2. Make ballpark estimates. Write a number model for each estimate.

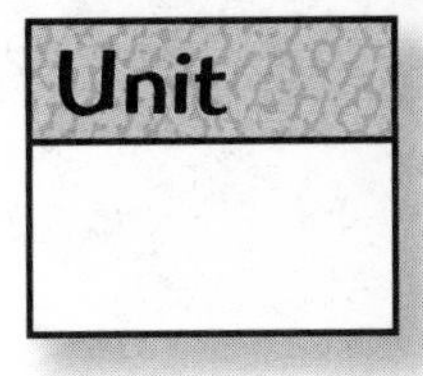

73 − 49

_____ − _____ = _____

87 − 21

_____ − _____ = _____

3. Match dimes to ¢.

5 dimes	100¢
10 dimes	150¢
15 dimes	50¢
18 dimes	180¢

4. Use the partial-sums algorithm to solve. Show your work.

$$\begin{array}{r} 46 \\ +\ 35 \\ \hline \end{array}$$

5. Solve.

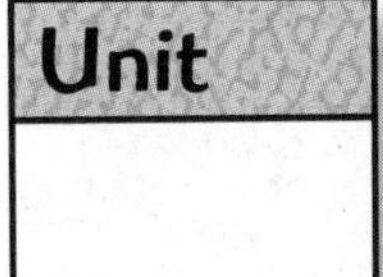

_____ = 6 + 5

16 + 5 = _____

46 + 5 = _____

_____ = 76 + 5

86 + 5 = _____

6. Write the names of 3 objects shaped like rectangular prisms.

____

____

____

Date Time

Math Boxes 5.10

1. Write the fact family.

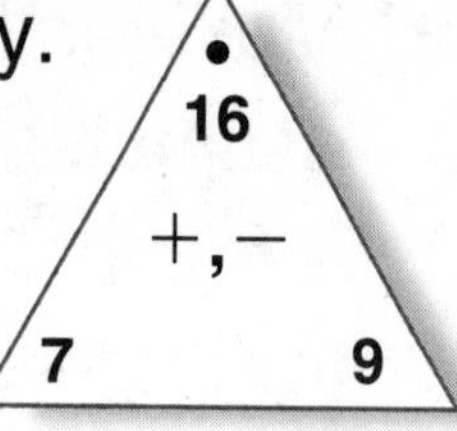

___ + ___ = ___

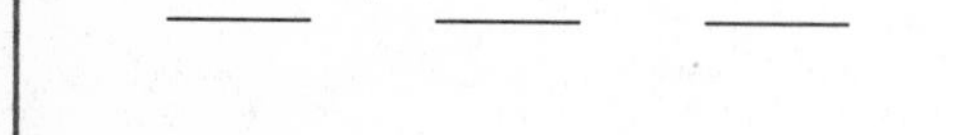

___ + ___ = ___

___ − ___ = ___

___ − ___ = ___

2. Draw a square. Make each side 3 cm long.

3. A triangle has _____ sides.

A rhombus has _____ sides.

A trapezoid has _____ sides.

A hexagon has _____ sides.

4. Use the partial-sums algorithm to solve. Show your work.

$$\begin{array}{r} 29 \\ +\ 53 \\ \hline \end{array}$$

5. How old will you be in 18 years?

_____ years old

Fill in the diagram and write a number model.

____

6. Write 3 numbers that add up to 20.

_____ + _____ + _____ = _____

Date Time

Three Addends

Materials ☐ number cards 0–20

Players 2

Directions

- Shuffle the cards. Place the deck number-side down.
- Turn over the top 3 cards. Each partner writes the 3 numbers.
- Add the numbers. Write a number model to show the order in which you added.
- Compare your answers with your partner's.

Example

The cards 6, 5, and 14 are turned over. Gillian records the numbers. She adds 14 and 6 first and then adds 5. She records her number model and compares her answer with her partner's.

Numbers: 6, 5, 14 Number model: 14 + 6 + 5 = 25

1. Numbers: _____, _____, _____

Number model:

_____ + _____ + _____ = _____

2. Numbers: _____, _____, _____

Number model:

_____ + _____ + _____ = _____

3. Numbers: _____, _____, _____

Number model:

_____ = _____ + _____ + _____

4. Numbers: _____, _____, _____

Number model:

_____ = _____ + _____ + _____

5. Numbers: _____, _____, _____

Number model:

_____ + _____ + _____ = _____

6. Numbers: _____, _____, _____

Number model:

_____ + _____ + _____ = _____

Date Time

Addition Practice

Fill in the unit box. Then, for each problem:

- Make a ballpark estimate before you add.
- Write a number model for your estimate.
- If your estimate is less than 50, you do not have to add the numbers. Leave the answer box empty.
- If your estimate is 50 or more, add the numbers. Write your answer in the answer box.

Unit

1. 29 + 7 **Answer** ______ Ballpark estimate: ______	**2.** 87 + 9 **Answer** ______ Ballpark estimate: ______	**3.** 37 + 42 **Answer** ______ Ballpark estimate: ______
4. 27 + 13 **Answer** ______ Ballpark estimate: ______	**5.** 38 + 46 **Answer** ______ Ballpark estimate: ______	**6.** 42 + 28 **Answer** ______ Ballpark estimate: ______

Date Time

Math Boxes 6.1

1. Write 3 numbers that add up to 19.

____ + ____ + ____ = ____

2. Complete each number model.

Unit

____ > 199

$372 >$ ____

____ < 424

$269 <$ ____

3. Use your calculator. Enter 37. Change to 67.

What did you do? ____

Enter 24. Change to 84.

What did you do? ____

4. Find the differences.

16°C and 28°C ____

70°F and 57°F ____

15°C and 43°C ____

5. Write the fact family.

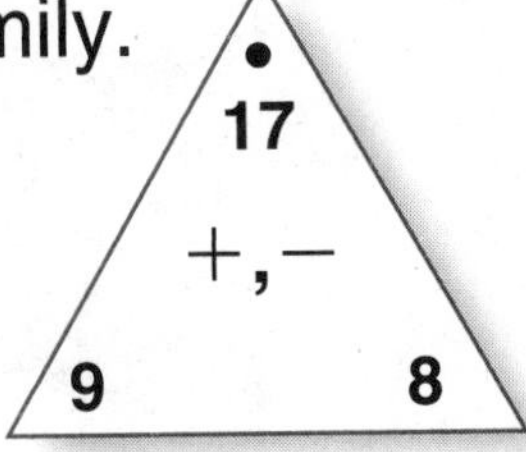

____ + ____ = ____

____ + ____ = ____

____ − ____ = ____

____ − ____ = ____

6. Use the partial-sums algorithm to solve. Show your work.

$$\begin{array}{r} 59 \\ +\ 32 \\ \hline \end{array}$$

Date Time

Comparison Number Stories

For each number story:

- Write the numbers you know in the comparison diagram.
- Write ? for the number you want to find.
- Solve the problem.
- Write a number model.

1. Barb scored 27 points.
Cindy scored 10 points.

Barb scored ______ more points than Cindy.

Number model: ____________________

Quantity	
Quantity	**Difference**

2. Jack scored 13 points.
Jack scored 6 more points than Eli.

Eli scored ______ points.

Number model: ____________________

Quantity	
Quantity	**Difference**

3. Frisky lives on the 16th floor.
Fido lives on the 7th floor.

Frisky lives ______ floors higher than Fido.

Number model: ____________________

Quantity	
Quantity	**Difference**

Date Time

Comparison Number Stories (cont.)

4. A jacket costs $75.
Pants cost $20.

The pants cost $_____ less than the jacket.

Number model: ________________

Quantity

Quantity	Difference

5. Ida is 36 years old.
Bob is 20 years old.

Ida is _____ years older than Bob.

Number model: ________________

Quantity

Quantity	Difference

6. Billy is 16 years old.
Paul is 6 years younger than Billy.

Paul is _____ years old.

Number model: ________________

Quantity

Quantity	Difference

7. Marcie is 56 inches tall.
Nick is 70 inches tall.

Marcie is _____ inches shorter than Nick.

Number model: ________________

Quantity

Quantity	Difference

Date Time

Math Boxes 6.2

1. Use $<$, $>$, or $=$.

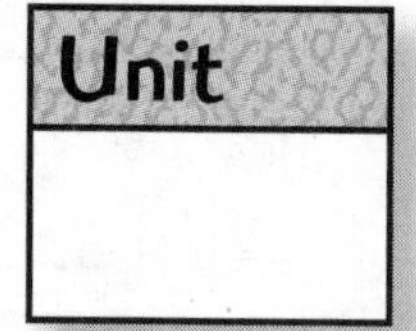

$7 + 5 + 30$ _____ 40

$11 + 6 + 4$ _____ 26

32 _____ $18 + 7 + 2$

19 _____ $13 + 9 + 1$

2. Write three names for 15.

_____ + _____ + _____ = 15

15 = _____ + _____ + _____

_____ + 9 + _____ = 15

3. Make a ballpark estimate. Write a number model for your estimate.

$49 + 51$

_____ + _____ = _____

4. Use your calculator. Enter 49. Change to 99.

What did you do? _________

Enter 15. Change to 85.

What did you do? _________

5. Tim harvested 12 bushels of corn and 19 bushels of tomatoes. How many bushels in all? _____ bushels

Fill in the diagram and write a number model.

Total	
Part	**Part**

____

6. Draw a line segment 8 cm long.

Draw a line segment 3 cm shorter than the one you just drew.

Date Time

What Is Your Favorite Food?

1. Make tally marks to show the number of children who chose a favorite food in each group.

fruit/vegetables	bread/cereal/ rice/pasta	dairy products	meat/poultry/fish/ beans/eggs/nuts

2. Make a graph that shows how many children chose a favorite food in each group.

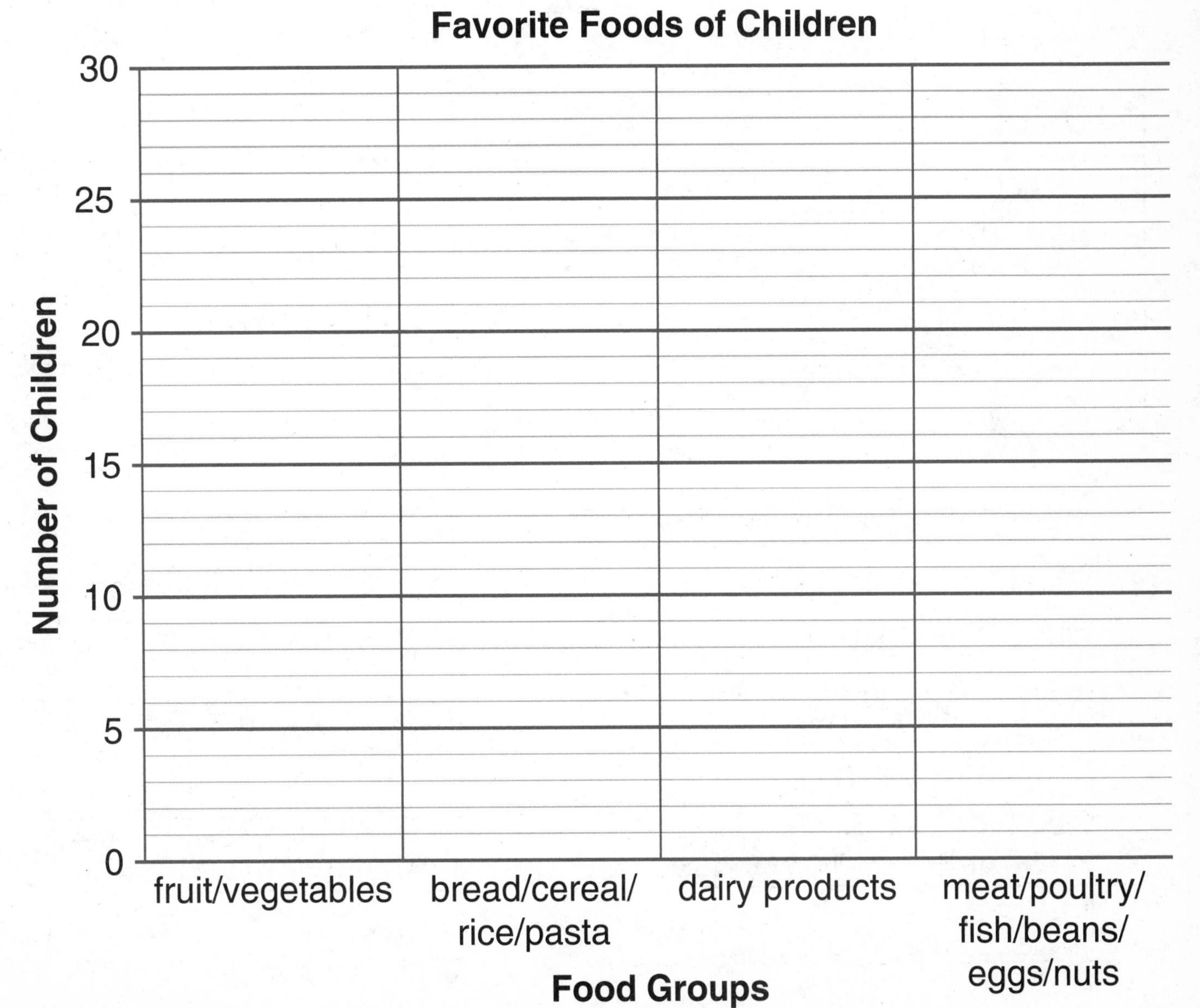

Date Time

Comparing Fish

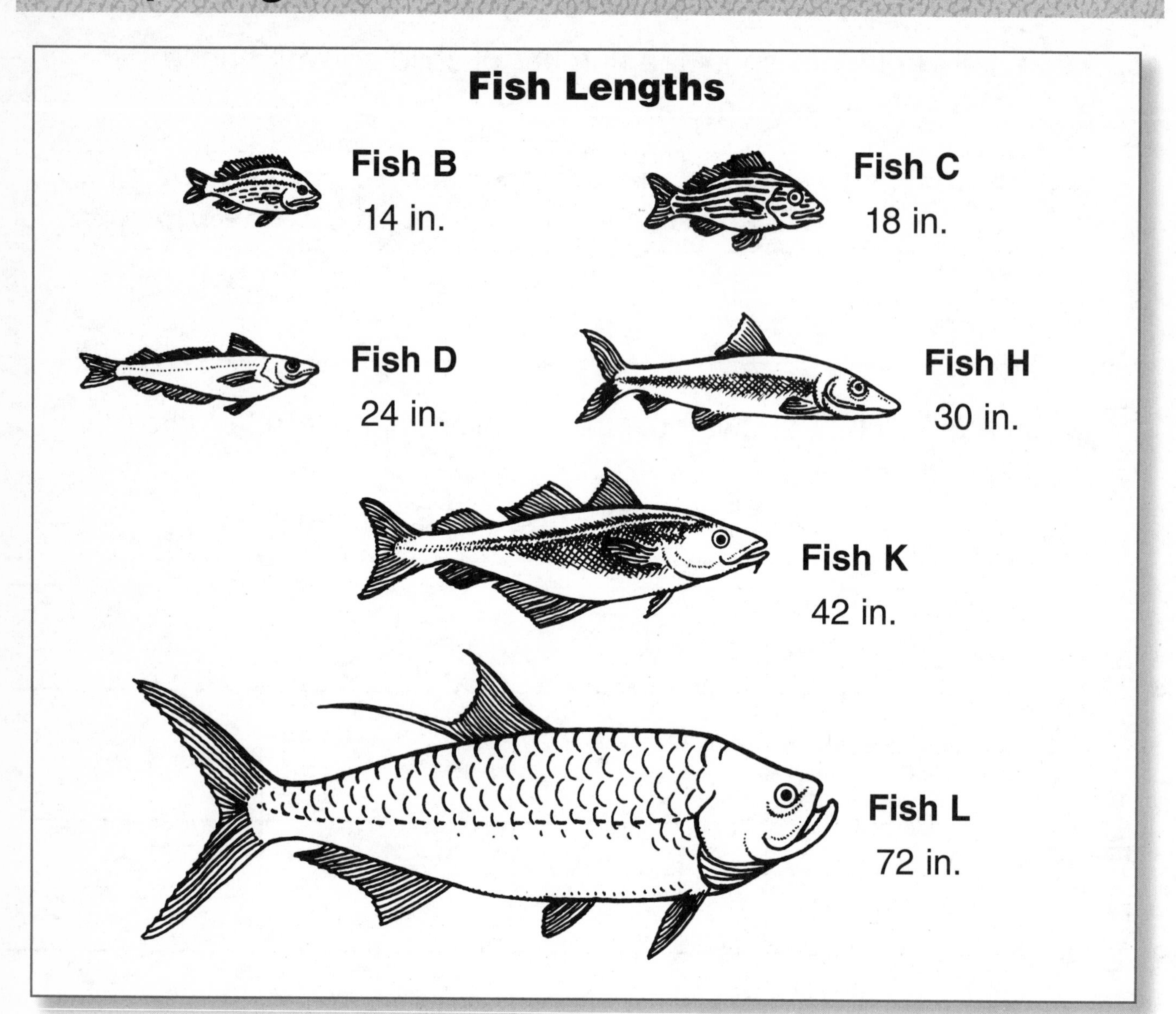

1. Fish C is _____ inches longer than Fish B.

Quantity
Fish C 18 in.

Quantity	Difference
Fish B 14 in.	

2. Fish H is _____ inches shorter than Fish K.

Quantity
Fish K ____

Quantity	Difference
Fish H ____	

Use with Lesson 6.3.

Date Time

Comparing Fish (cont.)

3. Fish L is ______ inches longer than Fish H.

Quantity

Quantity	Difference

4. Fish B is ______ inches shorter than Fish D.

Quantity

Quantity	Difference

5. Fish H is 6 inches longer than ______________.

Quantity

Quantity	Difference

6. Fish L is 30 inches longer than ______________.

Quantity

Quantity	Difference

7. Fish C is 6 inches shorter than ______________.

Quantity

Quantity	Difference

8. Fish B is 16 inches shorter than ______________.

Quantity

Quantity	Difference

Date Time

Math Boxes 6.3

1. 35 first graders. 48 second graders. How many more children in second grade?

____ more

Fill in the diagram and write a number model.

Quantity

Quantity	Difference

____

2. Solve.

Unit

30 + ____ = 64

73 = 20 + ____

____ + 50 = 82

55 = ____ + 10

3. Use the partial-sums algorithm to solve. Show your work.

$$\begin{array}{r} 27 \\ +\ 56 \\ \hline \end{array}$$

4. Complete each number model.

Unit

________ > 209

526 > ________

________ < 317

461 < ________

5. What is the temperature?

____

Is it warm or cold?

____

°C 30 20

6.

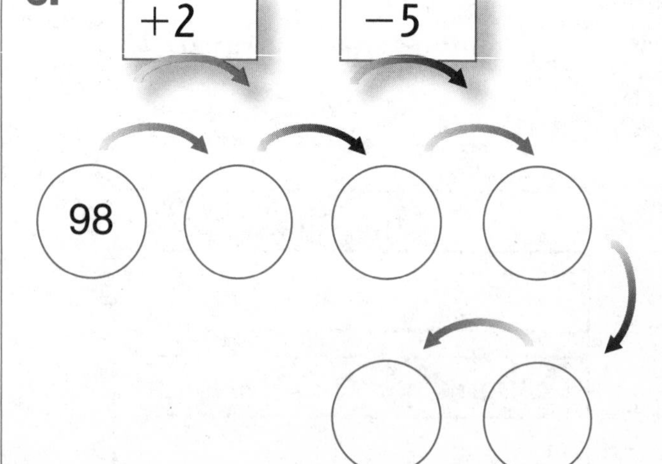

Date Time

Addition and Subtraction Number Stories

Do the following for each problem:

- Choose one diagram from *Math Masters,* page 108.
- Fill in the numbers in the diagram. Write ? for the number you want to find. Find the answer. Write a number model.
- In Problems 3 and 4, write your own unit.

1. Rushing Waters has 26 water slides. Last year, there were only 17 water slides. How many new slides are there this year?

There are ____ new water slides.

Number model:

____

2. The Loop Slide is 65 feet high. The Tower Slide is 48 feet high. How much shorter is the Tower Slide?

It is ____ feet shorter.

Number model:

____

3. Colin has 20 __________.

Fiona has 30 __________.

How many __________ do they have in all?

Colin and Fiona have

____ ________ (unit) in all.

Number model:

____

4. Alexi had 34 __________.

He gave 16 __________ to Theo. How many __________ does Alexi have now?

Alexi has ____ ________ (unit) now.

Number model:

____

Date Time

Math Boxes 6.4

1. 28 soccer balls. 36 basketballs. How many more basketballs?

_____ basketballs

Fill in the diagram and write a number model.

Quantity

Quantity	Difference

____

2. Write even or odd.

126 _______

311 _______

109 _______

430 _______

3. A toy costs $1.25. You pay with two dollar bills. How much change do you get?

______¢ or $______

4. Fill in the pieces of the grid.

		565	

5. What time is it?

_____ : _____

In 15 minutes it will be

_____ : _____.

6. 35 butterflies. 10 flew away. How many butterflies are left? _____ butterflies

Fill in the diagram and write a number model.

Quantity

Quantity	Difference

____

Date Time

Subtraction with Base-10 Blocks

Use base-10 blocks to help you subtract.

1.

longs 10s	cubes 1s
1	9
–	7

2.

longs 10s	cubes 1s
2	5
– 1	4

3.

longs 10s	cubes 1s
4	4
– 3	1

4.

longs 10s	cubes 1s
3	6
– 2	3

5.

longs 10s	cubes 1s
2	9
– 1	8

6.

longs 10s	cubes 1s
4	7
– 2	5

7.

longs 10s	cubes 1s
3	8
– 2	6

8.

longs 10s	cubes 1s
5	7
– 1	6

9.

longs 10s	cubes 1s
6	8
– 2	4

10.

longs 10s	cubes 1s
2	8
– 2	0

11.

longs 10s	cubes 1s
2	8
–	8

12.

longs 10s	cubes 1s
3	8
– 2	3

Use mental math to solve these problems:

13. 76 – 46 = ______

14. ______ = 37 – 10

15. 20 + ______ = 54

16. ______ + 30 = 97

Date Time

More Subtraction with Base-10 Blocks

Use base-10 blocks to help you subtract.

1.

longs 10s	cubes 1s
1	4
−	6

2.

longs 10s	cubes 1s
2	5
− 1	7

3.

longs 10s	cubes 1s
4	3
− 1	8

4.

longs 10s	cubes 1s
3	2
− 1	9

5.

longs 10s	cubes 1s
2	4
− 1	5

6.

longs 10s	cubes 1s
4	5
− 2	7

7.

longs 10s	cubes 1s
3	6
− 2	8

8.

longs 10s	cubes 1s
5	6
− 1	7

9.

longs 10s	cubes 1s
6	4
− 2	6

10.

longs 10s	cubes 1s
3	4
− 1	8

11.

longs 10s	cubes 1s
4	1
− 2	6

12.

longs 10s	cubes 1s
5	5
− 4	8

Use mental math to solve these problems:

13. 83 − 23 = ______

14. ______ = 46 − 20

15. 40 + ______ = 72

16. ______ + 20 = 93

Date Time

Diagram Problems

Which number will make the diagram correct? Fill in the oval next to the correct answer.

Example:
- ○ $T = 16$
- ● $T = 13$
- ○ $T = 3$

Total	
T	
Part	Part
8	5

1.
- ○ $A = 20$
- ○ $A = 63$
- ○ $A = 23$

Quantity	
43	
Quantity	Difference
20	A

2.
- ○ $B = 7$
- ○ $B = 25$
- ○ $B = 9$

Start	Change	End
16	−9	B

3.
- ○ $C = 120$
- ○ $C = 60$
- ○ $C = 40$

Total	
90	
Part	Part
30	C

4.
- ○ $D = 80$
- ○ $D = 20$
- ○ $D = 40$

Start	Change	End
D	+20	60

Date Time

Math Boxes 6.5

1. Use the partial-sums algorithm to solve. Show your work.

$$\begin{array}{r} 45 \\ +\ 38 \\ \hline \end{array}$$

2. Write three names for 18.

18 = _____ + _____ + _____

_____ + 10 + _____ = 18

18 = _____ + 5 + _____

3. What is your age now?

____

In 15 years, your age will be

_____________________.

4. Find the rule. Fill in the table.

Rule

in	out
368	358
229	219
541	
213	

5. Write 3 even numbers larger than 100.

____

____

____

6. Solve.

Unit

100 = 49 + _____

24 + _____ = 100

100 = 57 + _____

$$\begin{array}{r} 38 \\ +\ \square \\ \hline 100 \end{array}$$

Trade-First Subtraction

- Make a ballpark estimate for each problem and write a number model for your ballpark estimate.
- Use the trade-first method of subtraction to solve each problem.
- Check your answers at the bottom of journal page 151.

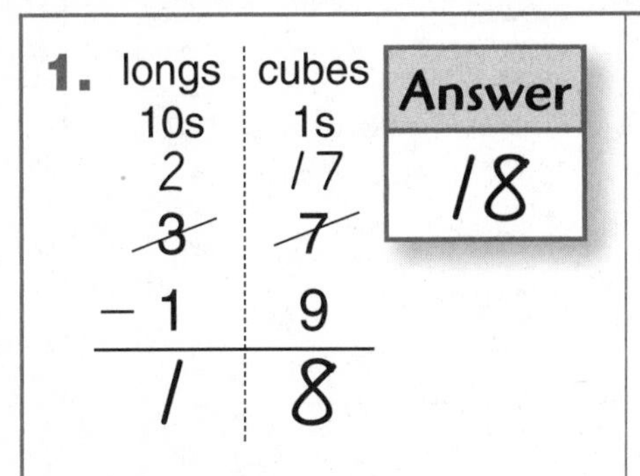

1.

longs 10s	cubes 1s
~~3~~ 2	~~7~~ 17
− 1	9
1	8

Answer: 18

Ballpark estimate: $40 - 20 = 20$

2.

longs 10s	cubes 1s
2	8
− 1	9

Answer:

Ballpark estimate:

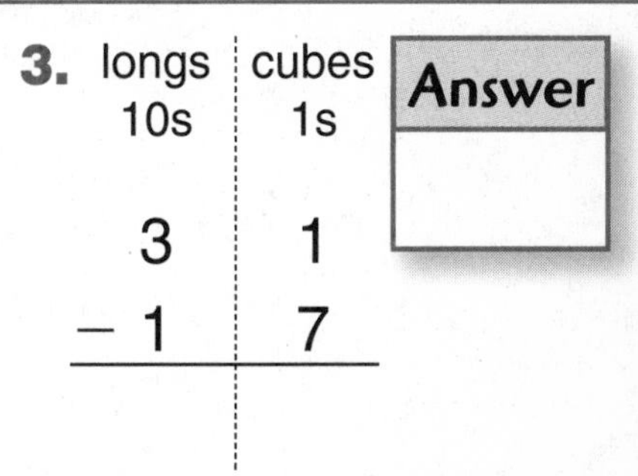

3.

longs 10s	cubes 1s
3	1
− 1	7

Answer:

Ballpark estimate:

4.

longs 10s	cubes 1s
5	8
− 3	4

Answer:

Ballpark estimate:

5.

longs 10s	cubes 1s
7	6
− 5	9

Answer:

Ballpark estimate:

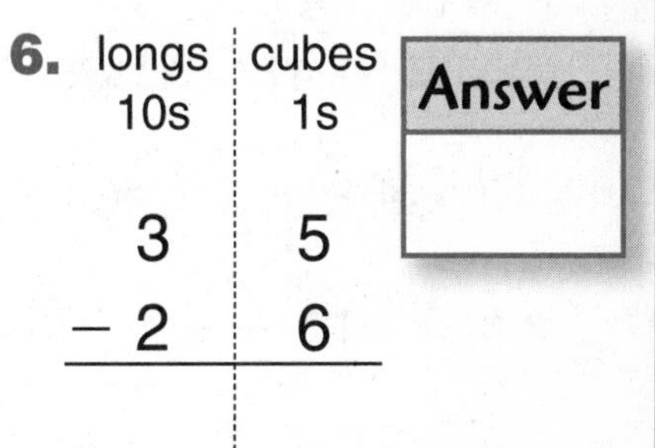

6.

longs 10s	cubes 1s
3	5
− 2	6

Answer:

Ballpark estimate:

7.

longs 10s	cubes 1s
8	3
− 4	2

Answer:

Ballpark estimate:

8.

longs 10s	cubes 1s
4	4
− 2	7

Answer:

Ballpark estimate:

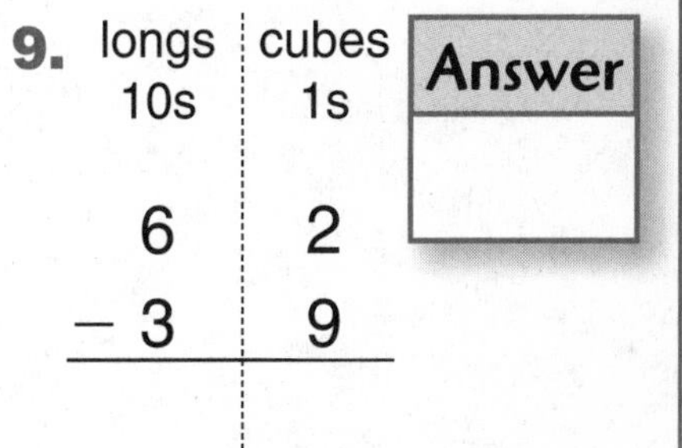

9.

longs 10s	cubes 1s
6	2
− 3	9

Answer:

Ballpark estimate:

Date Time

Math Boxes 6.6

1. Solve.

Unit
robots

72 = ____ + 60

58 = 30 + ____

____ + 70 = 81

40 + ____ = 94

2. Write 5 names for 150.

150

3.

How many cubes? ____

Cross out 13 cubes.

How many are left? ____

Write the number model.

____ – ____ = ____

4. Draw the hands to show 6:45.

$\frac{1}{2}$ hour earlier is ____ : ____.

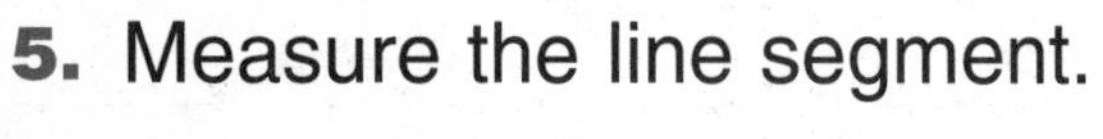

5. Measure the line segment.

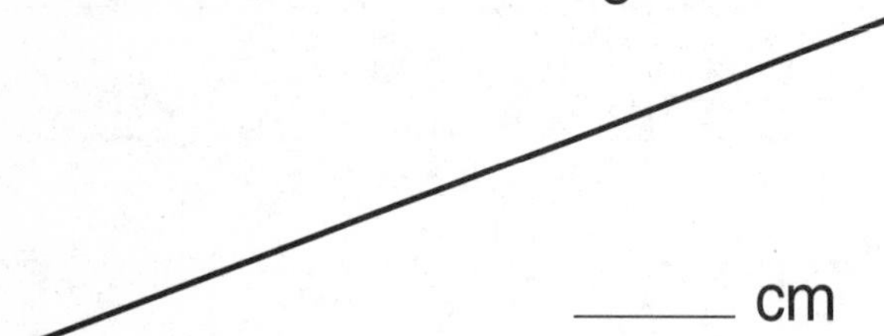

____ cm

Draw a line segment 3 cm shorter.

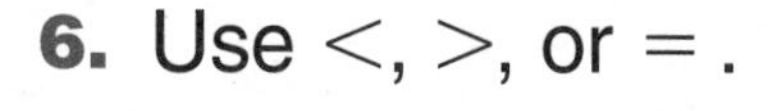

6. Use $<$, $>$, or $=$.

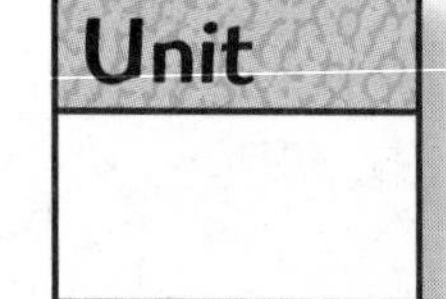

50 ____ 25 + 5 + 10

29 ____ 12 + 9 + 8

30 + 9 + 11 ____ 43

67 ____ 13 + 7 + 30

Date Time

How Many Persons Get *n* Things?

Follow the directions on *Math Masters,* page 114 to fill in the table.

What is the total number of counters?	**How many counters are there in each group?** (Roll a die to find out.)	**How many groups are there?**	**How many counters are left over?**

Answers for page 149:

1. 18 **2.** 9 **3.** 14

4. 24 **5.** 17 **6.** 9

7. 41 **8.** 17 **9.** 23

Math Boxes 6.7

1. 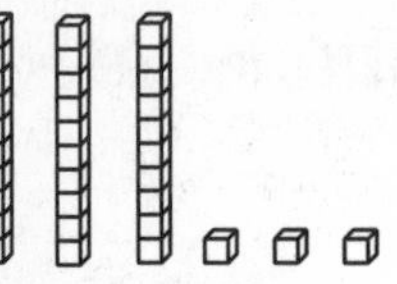

How many cubes? _____

Cross out 14 cubes.

How many are left? _____

Write the number model.

_____ – _____ = _____

2. Make a ballpark estimate. Write a number model.

$$\begin{array}{r} 48 \\ -\ 29 \\ \hline \end{array}$$

Ballpark estimate:

$$\begin{array}{r} \square \\ -\ \square \\ \hline \square \end{array}$$

Number model:

____

3. There are 22 first graders and 35 second graders. How many more second graders?

_____ more

Fill in the diagram and write a number model.

Quantity	
Quantity	**Difference**

____

4. 15 cats. 29 kittens were born. How many animals are there now? _____ animals

Fill in the diagram and write a number model.

Start	Change	End

____

5. How much money?

_______¢ or $_______

6. 14 bunnies, 5 birds, 6 frogs.

There are _______ animals in all.

Write a number model.

____

Date Time

Multiplication Stories

Solve each problem. Draw pictures or use counters to help.

Example: How many cans are in three 6-packs of juice?

/// /// ///
/// /// ///
6 12 18

Answer: 18 cans

1. Mr. Yung has 4 boxes of markers. There are 6 markers in each box. How many markers does he have in all?

Answer: ______ markers

2. Sandi has 3 bags of marbles. Each bag has 7 marbles in it. How many marbles does she have in all?

Answer: ______ marbles

3. Mrs. Jayne brought 5 packages of buns to the picnic. Each package had 6 buns in it. How many buns did she bring in all?

Answer: ______ buns

4. After the picnic, 5 boys each picked up 4 soft-drink cans to recycle. How many cans did the boys pick up altogether?

Answer: ______ cans

Math Boxes 6.8

1. Draw two polygons with 4 sides. Can you name your polygons?

_______________ _______________

2. Count by quarters. Start at $3.00.

$3.00, _______, _______,

_______, _______, _______,

_______, _______, $5.00

3. How many cubes? _____

Cross out 23 cubes.

How many are left? _____

Write a number model.

_____ – _____ = _____

4. Use the partial-sums algorithm to solve. Show your work.

$$\begin{array}{r} 36 \\ +\ 49 \\ \hline \end{array}$$

5. 26 red balls. 55 blue balls.

How many in all? _____ balls

Fill in the diagram and write a number model.

Total	
Part	Part

____

6. Solve.

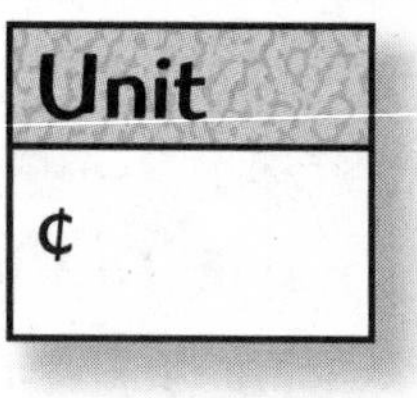

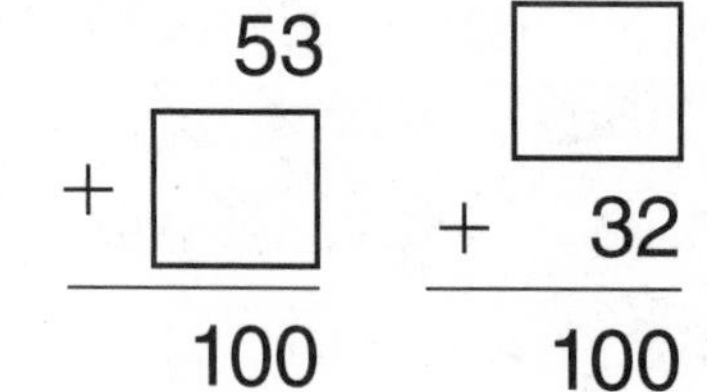

$$\begin{array}{r} 53 \\ +\ \square \\ \hline 100 \end{array} \qquad \begin{array}{r} \square \\ +\ 32 \\ \hline 100 \end{array}$$

$$\begin{array}{r} 46 \\ +\ \square \\ \hline 100 \end{array} \qquad \begin{array}{r} \square \\ +\ 88 \\ \hline 100 \end{array}$$

Date Time

Array Number Stories

Array

Multiplication Diagram

rows	____ per row	____ in all

Number model: ____ × ____ = ____

Array

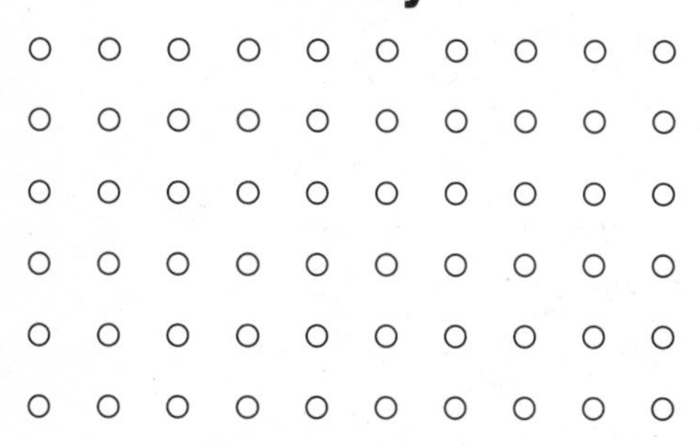

Multiplication Diagram

rows	____ per row	____ in all

Number model: ____ × ____ = ____

Array

Multiplication Diagram

rows	____ per row	____ in all

Number model: ____ × ____ = ____

Array

Multiplication Diagram

rows	____ per row	____ in all

Number model: ____ × ____ = ____

Number Stories

- Use Xs to show the array.
- Answer the question.
- Fill in the number model.

1. The marching band has 3 rows with 5 players in each row. How many players are in the band?

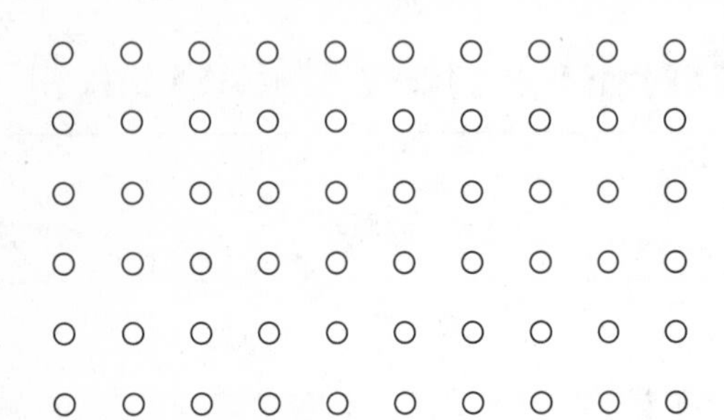

There are ______ players in the band.

______ × ______ = ______

2. The orchard has 4 rows of trees. Each row has 8 trees. How many trees are there?

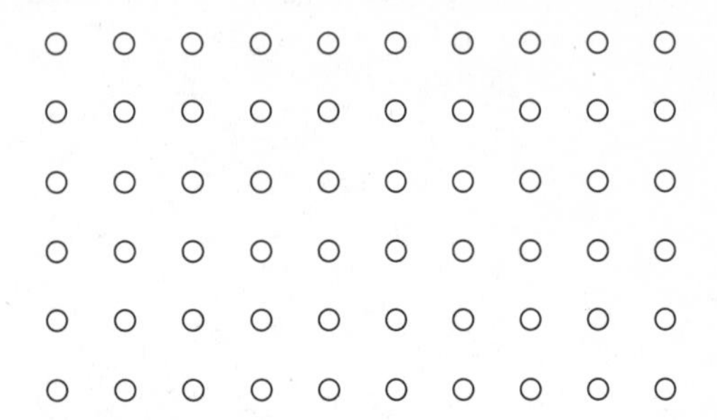

There are ______ trees in the orchard.

______ × ______ = ______

3. The sheet has 5 rows of stamps. There are 5 stamps in each row. How many stamps are there?

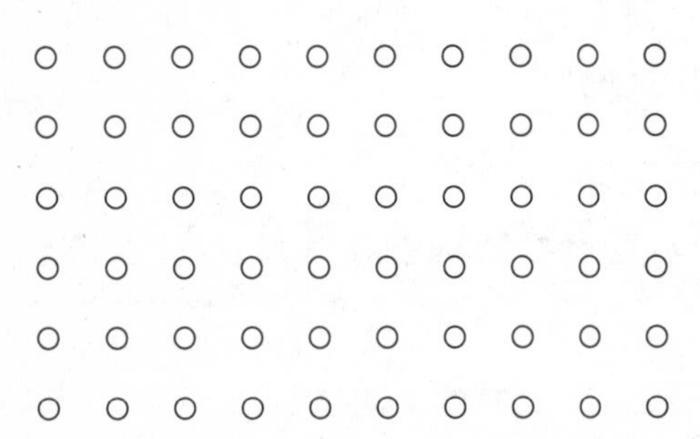

There are ______ stamps in all.

______ × ______ = ______

4. Mel folded his paper into 2 rows of 4 boxes each. How many boxes did he make?

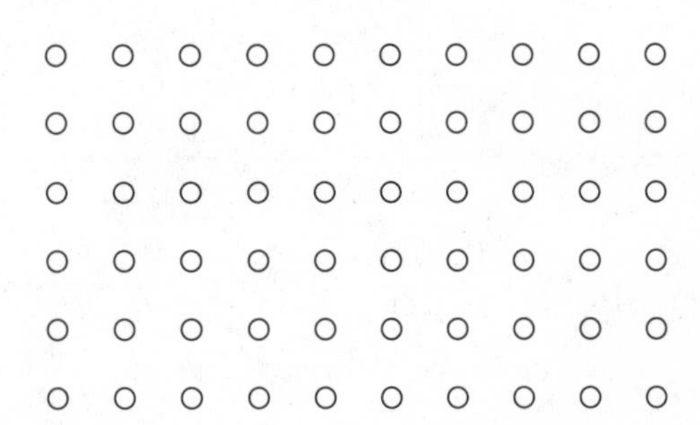

He made ______ boxes.

______ × ______ = ______

Date Time

More Multiplication Number Stories

Write your own multiplication stories and draw pictures of your stories. You can use the pictures at the side of the page for ideas.

For each story:

- Write the words.
- Draw a picture.
- Write the answer.

Example

There are 5 tricycles. How many wheels in all?

Answer: 15 wheels
(unit)

1. ______________________________

____

Answer: ______________
(unit)

2. ______________________________

____

Answer: ______________
(unit)

A person has 2 ears.

A tricycle has 3 wheels.

A car has 4 wheels.

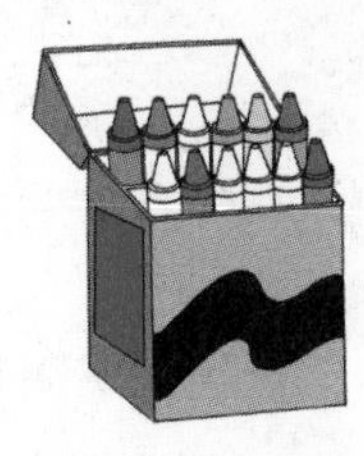

The box has 12 crayons.

The box has 100 paper clips.

The juice pack has 6 cans.

Date Time

Math Boxes 6.9

1. Follow the arrow rules. Fill in the missing frames.

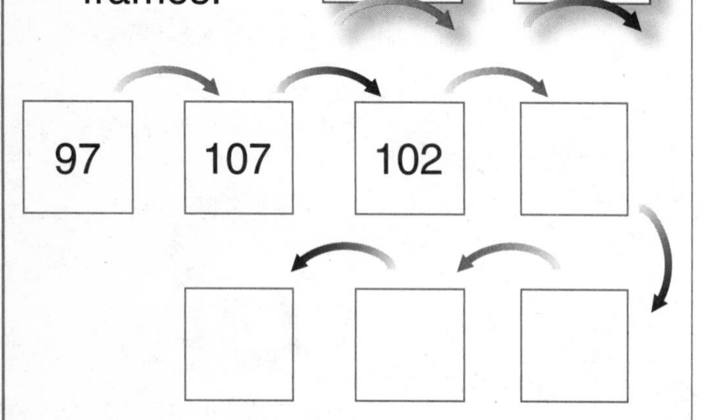

2. 5 hands. 5 fingers on each hand. How many fingers? Draw a picture.

There are ______ fingers.

3. Write a number model for a ballpark estimate. Then subtract to solve.

$$\begin{array}{r} 73 \\ -\ 41 \\ \hline \end{array}$$

Answer

Ballpark estimate:

____

4. 9 pigs, 5 cats, and 6 dogs. How many animals in all?

______ animals

Fill in the diagram. Write a number model.

Total		
Part	Part	Part

____

5. How many stars in all?

______ stars

Fill in the multiplication diagram.

rows	______ per row	______ in all

6. Use the partial-sums algorithm to solve. Show your work.

$$\begin{array}{r} 17 \\ +\ 64 \\ \hline \end{array}$$

Date Time

Trade-First Subtraction

For each problem, do the following:

- Make a ballpark estimate before you subtract. Write a number model for your estimate.
- If your estimate is less than 50, subtract the numbers. Write your answer in the answer box.
- If your estimate is 50 or more, you do not have to find an exact answer. Leave the answer box empty.

1. 68 − 43 Answer: ____ Ballpark estimate: ____	**2.** 56 − 39 Answer: ____ Ballpark estimate: ____	**3.** 73 − 14 Answer: ____ Ballpark estimate: ____
4. 47 − 19 Answer: ____ Ballpark estimate: ____	**5.** 88 − 23 Answer: ____ Ballpark estimate: ____	**6.** 82 − 65 Answer: ____ Ballpark estimate: ____
7. 91 − 26 Answer: ____ Ballpark estimate: ____	**8.** 94 − 18 Answer: ____ Ballpark estimate: ____	**9.** 64 − 38 Answer: ____ Ballpark estimate: ____

Date Time

Math Boxes 6.10

1. 4 rows of 4 chairs. How many chairs in all? _____ chairs

Draw an array to solve. Fill in the multiplication diagram.

rows	_____ per row	_____ in all

2. Solve.

Unit
¢

57 + _____ = 60

40 = 35 + _____

74 + _____ = 80

120 = 111 + _____

3. Draw a line segment 3 inches long.

Now draw a line segment 1 inch shorter.

4. Follow the rule. Fill in the table.

Rule
+11

in	out
73	
159	
	96

5. Write 5 names for 28.

28

6. Sue has $1.00 and spends 73¢. How much change does she get?

_____

Date Time

Division Problems

Use counters or simple drawings to find the answers. Fill in the blanks.

1. 16 cents shared equally

by 2 people

_____¢ per person

_____¢ remaining

by 4 people

_____¢ per person

_____¢ remaining

by 3 people

_____¢ per person

_____¢ remaining

by 5 people

_____¢ per person

_____¢ remaining

2. 25 cents shared equally

by 3 people

_____¢ per person

_____¢ remaining

by 4 people

_____¢ per person

_____¢ remaining

by 5 people

_____¢ per person

_____¢ remaining

by 6 people

_____¢ per person

_____¢ remaining

3. 16 crayons, 6 crayons per box

How many boxes? _____ How many crayons remaining? _____

4. 24 eggs, 6 eggs in each cake

How many cakes? _____ How many eggs remaining? _____

Date Time

Math Boxes 6.11

1. Fill in the missing numbers.

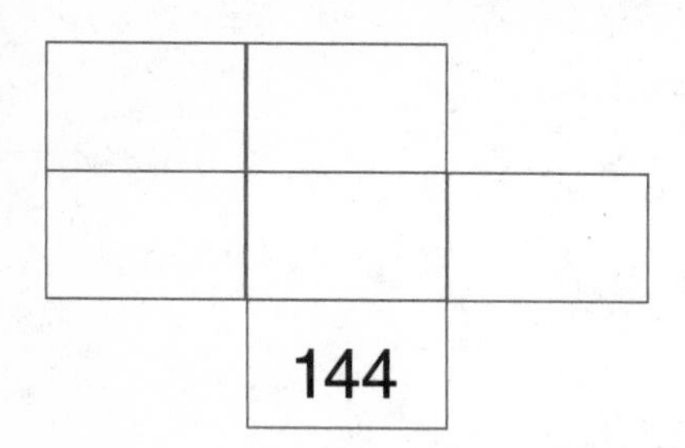

144

2. Write the fact family.

____ + ____ = ____

____ + ____ = ____

____ − ____ = ____

____ − ____ = ____

3. Write a number model for a ballpark estimate. Then subtract to solve.

$$\begin{array}{r} 66 \\ -\ 47 \\ \hline \end{array}$$

Ballpark estimate:

____

4. The temperature is

____ °F.

°F

60

50

5. Continue.

88, 98, ____, ____,

____, ____, ____,

____, ____

6. Name 3 objects shaped like a cone.

____

____

____

Date Time

Math Boxes 6.12

1. 4 children share 12 slices of pizza equally. How many slices does each child get? Draw a picture.

   Each child gets _____ slices.

2. Fill in the missing number on the Fact Triangle. Then write the fact family.

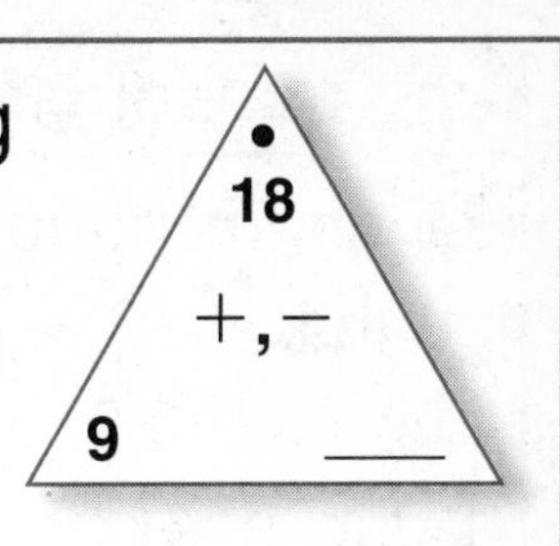

   _____ + _____ = _____

   _____ − _____ = _____

3. Share 18¢ equally among 5 children. How many cents does each child get?

   _____

   How many cents are left over?

   _____

4. Solve.

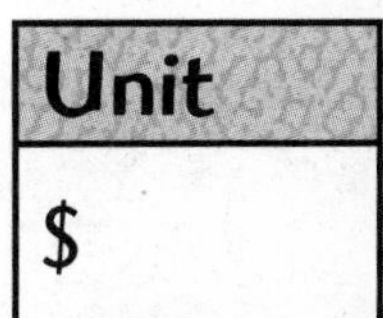

   5 + 6 + 23 = _____

   _____ = 3 + 3 + 12

   4 + 3 + 17 = _____

   _____ = 9 + 2 + 9

5. How many dots are in this 4-by-6 array? Count by 4s.

   _____ dots in all

6. Count by 2s.

   70, 72, 74, _____, _____,

   _____, _____, _____,

   _____, _____, _____,

   _____

Date Time

Table of Equivalencies

Weight	
kilogram:	1,000 g
pound:	16 oz
ton:	2,000 lb
1 ounce is about 30 g	

$<$	*is less than*
$>$	*is more than*
$=$	*is equal to*
$=$	*is the same as*

Length	
kilometer:	1,000 m
meter:	100 cm or 10 dm
decimeter:	10 cm
centimeter:	10 mm
foot:	12 in.
yard:	3 ft or 36 in.
mile:	5,280 ft or 1,760 yd
10 cm is about 4 in.	

Time	
year:	365 or 366 days
year:	about 52 weeks
year:	12 months
month:	28, 29, 30, or 31 days
week:	7 days
day:	24 hours
hour:	60 minutes
minute:	60 seconds

Money		
	1¢, or $0.01	
	5¢, or $0.05	
	10¢, or $0.10	
	25¢, or $0.25	
	100¢, or $1.00	$1

Capacity
1 pint = 2 cups
1 quart = 2 pints
1 gallon = 4 quarts
1 liter = 1,000 milliliters

Abbreviations	
kilometers	km
meters	m
centimeters	cm
miles	mi
feet	ft
yards	yd
inches	in.
tons	T
pounds	lb
ounces	oz
kilograms	kg
grams	g
decimeters	dm
millimeters	mm
pints	pt
quarts	qt
gallons	gal
liters	L
milliliters	mL

Date Time

Notes

Date Time

Notes

Date Time

Notes

Date Time

Notes

Date Time

+, − Fact Triangles 1

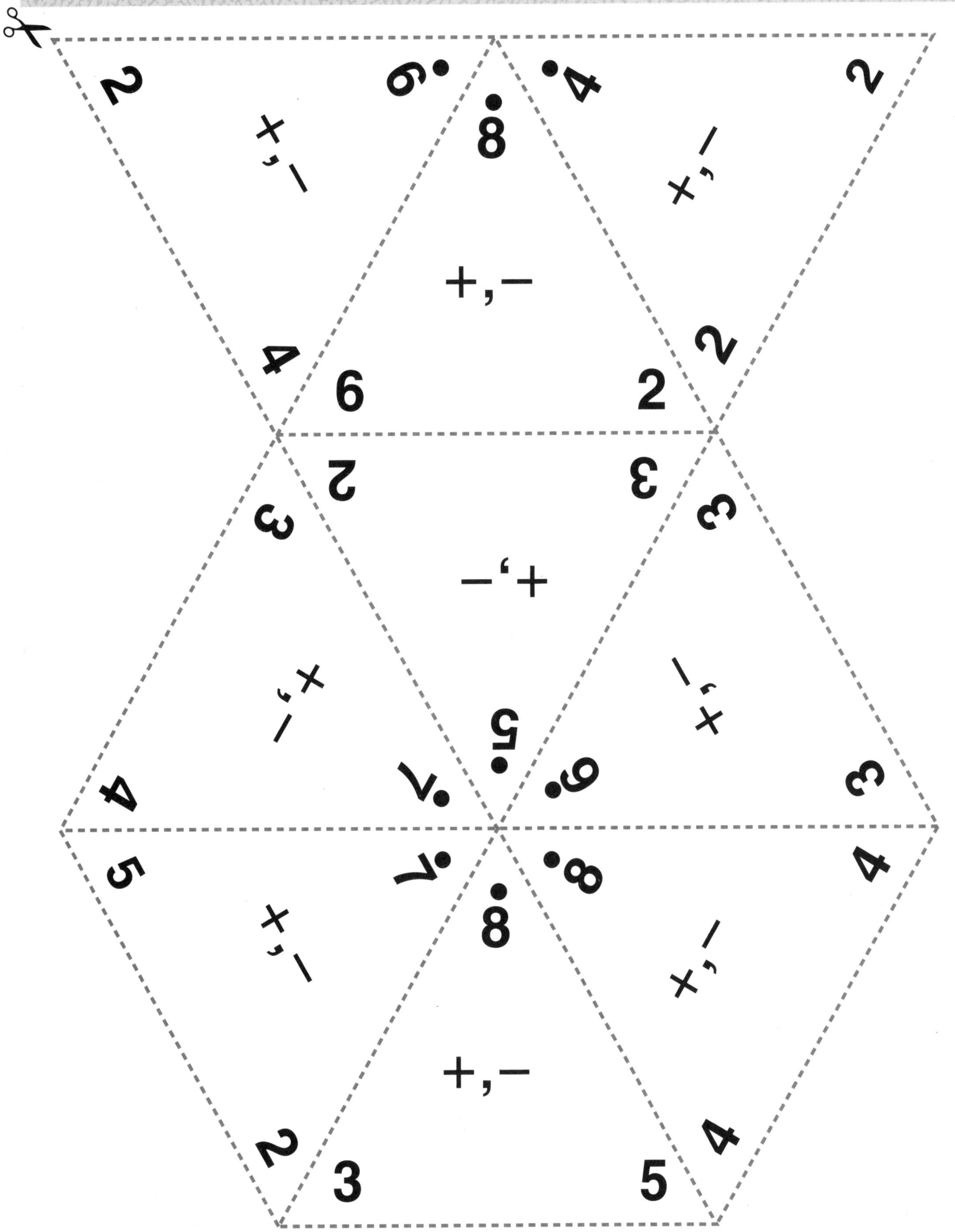

Date Time

+, − Fact Triangles 2

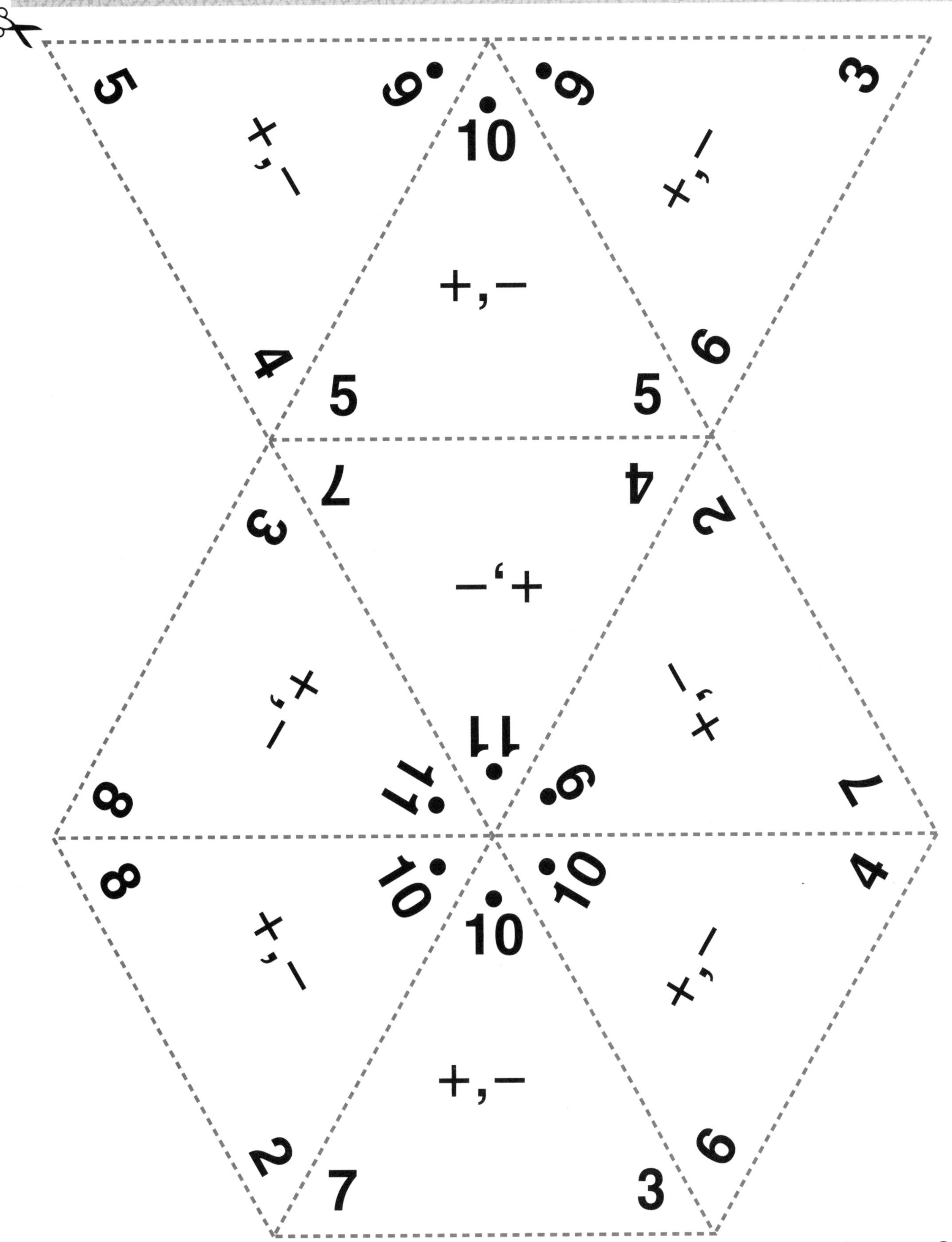